Tinkering

Kids Learn by Making Stuff

Curt Gabrielson

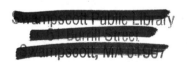

SAN FRANCISCO, CA

To Paulo and Zoraya. Long may you tinker!

Contents

Foreward

Tom Wolfe wrote a feature article in the December 1983 issue of Esquire magazine called "The Tinkerings of Robert Noyce." Wolfe tells the very American story of a young man who grew up in Grinnell, Iowa, where he went to college before going on to MIT for graduate school. After school, Noyce headed to California in 1956 where he would invent the electronic future as a co-founder of Intel and shape what we now call the Silicon Valley.

Wolfe points out that Noyce had a typical Midwestern upbringing. He was a curious boy and a good athlete. When he was 13, he and his brothers read an article in *Popular Science* about a box kite that could lift a person off the ground. Noyce and his brothers set out to build and test that kite, asking themselves: would it work as they say? The boys would persist after several failures to get the kite up in the air. While Noyce was a good student, he almost got thrown out of college because of a prank. Fortunately, a teacher recognized Noyce's talent and stepped in to help. That teacher introduced Noyce to transistors, while few others had even heard of them, and Noyce was curious enough to wonder how they might be used.

Wolfe wonders why a generation of great engineers and scientists came from such unexpected places. "Just why was it that small-town boys from the Middle West dominated the engineering frontiers? Noyce concluded it was because in a small town you became a technician, a tinker, an engineer, and an inventor, by necessity. "In a small town," Noyce liked to say, "when something breaks down, you don't wait around for a new part, because it's not coming. You make it yourself." Noyce was fortunate to have two kinds of education: informal as well as formal. Growing up, he learned a lot outside of school, as did others who grew up on farms and in families that knew how to use tools and how to fix machines. Formal learning often doesn't make sense without informal learning. It offers too much theory without enough grounding in practice. Tinkering represents this kind of practical education that is often undervalued in formal settings.

Tinkering is not a field like chemistry or physics, yet it is worthy of study, particularly by those who want to engage kids as makers today. Tinkering is to making as running is to sports, as tapping your foot is to music. Tinkering is a process. It is an attitude. It is the means to fix, make, change, modify, and customize the world.

Curt Gabrielson and his colleagues at the Watsonville Environmental Science Workshop are pioneers in informal education. They are skilled practitioners, thoughtfully organizing learning experiences for children in a supportive context outside of school. In this book, Gabrielson shows how to create these meaningful experiences for students and how adults can be effective as facilitators of learning. Tinkering can help children build confidence in their own capabilities and explore the world they live in. All children deserve to have these opportunities, early and often, whether at home or even in school. What's more, I believe that today's children are demanding such learning experiences because they know how essential it is for them to grow as learners and become creative contributors to society. Like Noyce, many of them might already realize that you can't just buy what exists but instead "you have to make it yourself."

Think what it means to introduce more children to tinkering—more girls, more kids from different economic and ethnic backgrounds, more kids with different learning abilities, middle class kids who are bored in school and more middle-aged adults? If we can get more of us tinkering, who knows what tough problems we can solve, what discoveries we will find and what new things we will create?

—Dale Dougherty, 2013

Preface

You must have seen infants and toddlers tinkering with things: they'll focus intently on some little object, say a little wooden box, grasp it clumsily, claw at it, look at it from all sides, shake it, pound on it trying to break it open. At some point, all of a sudden—pop—it comes open! Now the undexterous hands gather the little animal figures lying nearby. One by one the hands jam figures into the box. More and more are stuffed in until they're sticking out the top. The lid is tried and found not to fit. It is mashed a few times, and then alternatives are attempted. More force is applied with large, local objects. A few figures are removed. The lid is tried again. An iterative process is begun until the lid finally snaps shut on the animals.

The kid just learned a bit about volume, three-dimensional space, opening and shutting, space and matter, properties of materials, arrangement of objects ("packing"), and the value of repeated attempts. This was done with no institution, no teacher, no curriculum, no explicit or predefined goals, no grades, no tests nor evaluation of any kind, no threat of punishment nor penalty, no reward, no praise, no scaffolding, no final discussion nor debriefing nor facilitated closure, but with immense and easily visible satisfaction. Young kids also learn like this in groups, with the significant advantage that they can learn from one another's input as well.

I've never once heard anyone ask of this situation, "Well, yes, she is having fun, but is she learning anything?" Everyone believes she is, and a multibillion-dollar industry is built on selling parents sophisticated versions of that box and those animals precisely because parents want their kid to learn in this way. On the contrary, and quite perplexingly, when older students are exploring and tinkering in just the same manner, especially if it happens to be in an institution mandated to carry out education, one can hardly describe the scene without a chorus of glowering skeptics chiming in, as if on cue: "Well, yes, they're having fun, but are they learning anything?"

Here's my answer, the answer of this book: heck yes they're learning something, and it may be the most valuable thing they've learned all week, and it may raise all sorts of questions in their minds that inspire them to learn more about what they're tinkering with, and it may start them on a path to a satisfying career, not to mention good fun on their own time, and it may put them in the driver's seat of their own education by realizing their competence and ability to learn through tinkering, and they may begin to demand more of just this sort of learning opportunity.

If you question how I know this learning took place in the course of that tinkering, I'll have to confide that I have no proof beyond the following: most kids have learned oodles and oodles of stuff, including talking and walking, texting, and skateboarding, with just this hit-and-miss, trial-and-success, seat-of-the-pants approach. I believe this is called "proof by inspection."

Now, you can get a PhD trying to show, incontrovertibly, that learning is happening in a tinkering environment, or attempting to work out exactly how it is happening. I'll certainly not stand in your way. That's far and away more important than developing the next generation of fill-in-the-bubble exams. But I'm not so interested in that. I'm comfortable with my gut instinct, and I'm enormously interested in and committed to trying to get more kids tinkering.

One of the great challenges scientists face when doing science is to hold their own experience as a single data point, worth no more and no less than the thousands of others they'll need to draw a legitimate conclusion. But as you tread the path toward drawing a conclusion on this topic, allow me to present to you my data point to add to yours and all those others.

I grew up tinkering. Some of my earliest memories are tinkering. I have some memories of my dad tinkering beside me, and many memories of him trying to explain questions that arose during my tinkering, but mostly my memories are of hours and hours of lone tinkering, and then hours and hours of tinkering with my nerdy little buddies. These are deliciously sweet memories, and, given time, I can detail hundreds of concepts and truths I uncovered in the course of that tinkering, some of which were life's absolute essentials and some of which I continue to use on a daily basis. I repeat: I learned that stuff through tinkering and because of tinkering.

I feel fortunate to have had this experience, and I see that it is often hard for people who have not grown up tinkering to learn to learn through tinkering (that was not a typo). It is certainly not impossible, but it's as challenging as taking up music after ignoring it for the first several decades of your life. That said, you can absolutely make great music, and more importantly fulfill your life by making music, even if you take it up at a late date. The same is true for tinkering. Thus I encourage you, that is, give courage to you, even if you have never in your memory tinkered[1], to begin tinkering today and don't look back.

I've been running tinkering programs through the Watsonville Environmental Science Workshop (WESW) for the last 15 years. Our staff of around 10 adults and 20 high-school helpers serves the Watsonville community both in schools and in afterschool sites. It never fails to stun me temporarily when a teacher or after-school facilitator has observed kids tinkering in our program and proclaims, as if offering breaking news, "The kids sure love it!" It is sad to think that perhaps it is not the norm but rather something rare and special to see joyful kids learning. I certainly see joy and enthusiasm as the status quo in all our programs. It's a rare day when nobody present has a eureka moment. We see that many kids hunger for this stuff, stay engaged a long time, stay excited about it long after they've walked away with their new creation, and even bug us about it later when we see them again.

We work with a lot of "at-risk" kids, rough kids who are falling through the cracks of the system. We actually seek out these kids because we see that they are often quite successful at tinkering, using tools, making a project work, and adding new ideas. It is clear that often this success is new to them and that it builds confidence. This sort of confidence is solid—not the ephemeral sort that comes and goes with an authority figure's praise—for after a successful tinkering experience, there is no question of the student's capabilities. She needs no external indication beyond the functional project in hand to know

that she has mastered those tools and materials and amassed those competencies, which are not soon lost. We also work with serious high-school students on track to high-powered colleges. They may know that

$$\vec{F} = q\vec{v} \times \vec{B}$$

but they've never seen nor felt that magnetic force (F) spin a tiny coil of thin wire, having an electric current (qv), and suspended between the poles of a normal battery in the magnetic field (B) of a small magnet. It's a working motor, merrily whirring away at around 600 rpm. We do this project even with elementary school students, but the high-school students can really sink their teeth into the concepts behind it.

Whether you're a sage tinkerer or just about to take the plunge, I bid you Godspeed in your plans to do tinkering with your students, kids. You must know it's much easier to sit and tell them stories, read from the textbook, and hand out questions and answers all prepackaged and indisputable. They won't gain much from that, though, and it won't feed their souls. They'll get much more out of exploring and creating with an open-ended objective and a variety of materials and tools.

How to use this book

I wrote this book for adults to read, but if you're a kid, welcome! Nothing here will cause you irreparable damage, and you may even get some insight into your own education. If you're an adult, you should know that in the project sections I'll be addressing you as the original student. This means I trust that you'll go have an authentic tinkering experience for yourself before you try to set one up for your students. Even if you do read the activities here, run out of time, and end up tinkering together with students without first trying it out solo, know that you'll be learning, just like them, and the more conscious you can be of this learning, the better you'll be able to facilitate theirs. This also means in the project sections, I'll tell you personally how to do it, but not much about how to lay it on your students.

---

1.   Though please, talk to your parents about those first two years. I'm pretty sure...

I try to avoid the "once removed" language of some education books, where the teacher/facilitator is magically considered to have a deep understanding of something that is new to them. Thus, I don't say, "tell the students the following" or "ask the students what they observe." Instead I'll tell you what to do and ask you what you observe. Then you can do the same with your students. In the text chapters I try to give you what tools and perspectives you need to make great tinkering happen on your own terms with your own group of students or your own kids, who you know much better than I. I wrote the book to be practical, first and foremost. I'll cite a few research findings and other info sources, but only when they apply straightaway to carrying out good teaching with tinkering.

This is something of a reference book, so feel free to scan down to the given topic you're concerned with. What will be presented comes primarily from what we've learned in our successful programming of over 20 years at the Community Science Workshops (CSWs) I've been part of in California, especially the Watsonville Environmental Science Workshop, a small arm of the City of Watsonville Department of Public Works. We've received a lot of valuable feedback on the serious tinkering we do, and we've used that to hone our programs to great effectiveness.

In the chapters that follow I'll give you a bit of background on why I think learning through tinkering is important now and has been throughout history (Chapter 2). I'll lay out what a good tinkering session looks like (Chapter 4). There are a whole passel of logistics involved in carrying out tinkering with your students, but don't worry, I've got you covered (Chapter 6). When teaching kids through tinkering it is best to think about the community or communities you're working in as well as the kids' roles in the little tinkering community you create (Chapter 8). I can assure you that many questions will arise in a good tinkering session, and you are not likely to know the answers to all of them, but that's OK (Chapter 10). Finally, as long as you're doing tinkering, you may as well align it to your state science standards and think about how you will assess the students (Chapter 12).

Now let's get to it! We'll start with a real tinkering activity!

Conventions Used in this Book

This icon indicates a tip, suggestion, or general note.

This icon indicates a caution or warning.

Safari® Books Online

Safari Books Online is an on-demand digital library that delivers expert content in both book and video form from the world's leading authors in technology and business.

Technology professionals, software developers, web designers, and business and creative professionals use Safari Books Online as their primary resource for research, problem solving, learning, and certification training.

Safari Books Online offers a range of plans and pricing for enterprise, government, education, and individuals. Members have access to thousands of books, training videos, and prepublication manuscripts in one fully searchable database from publishers like Maker Media, O'Reilly Media, Prentice Hall Professional, Addison-Wesley Professional, Microsoft Press, Sams, Que, Peachpit Press, Focal Press, Cisco Press, John Wiley & Sons, Syngress, Morgan Kaufmann, IBM Redbooks, Packt, Adobe Press, FT Press, Apress, Manning, New Riders, McGraw-Hill, Jones & Bartlett, Course Technology, and hundreds more. For more information about Safari Books Online, please visit us online.

How to
Contact Us

Please address comments and questions concerning this book to the publisher:

Make:
1160 Battery Street East, Suite 125
San Francisco, CA 94111
877-306-6253 (in the United States or Canada)
707-639-1355 (international or local)

We have a web page for this book, where we list errata, examples, and any additional information. You can access this page at *http://oreil.ly/make_tinkering*.

To comment or ask technical questions about this book, send email to *bookquestions@oreilly.com*.

Make: unites, inspires, informs, and entertains a growing community of resourceful people who undertake amazing projects in their backyards, basements, and garages. Make: celebrates your right to tweak, hack, and bend any technology to your will. The Make: audience continues to be a growing culture and community that believes in bettering ourselves, our environment, our educational system—our entire world. This is much more than an audience; it's a worldwide movement that Make: is leading—we call it the Maker Movement.

For more information about Make:, visit us online:

Make: magazine: http://makezine.com/magazine
Maker Faire: http://makerfaire.com
Makezine.com: http://makezine.com
Maker Shed: http://makershed.com

Acknowledgments

Thanks first to Gustavo Hernandez (and on and on), champion maker-of-all-things, tinkerer extraordinaire, soul mate and ultimate partner in crime for so many good years at the WESW. Thanks to Angelica Gonzalez who took over from me, and now the amazing Jose Sandoval, born and raised on tinkering at the Fresno Science Workshop. Thanks to all the WESW staff: Araceli Ortiz, Aurora Torres, Darren Gertler, Emilyn Green, Fabiola Pizano, Nestor Orozco, Omar Vigil, and Sal Lua, who make it happen for so many kids. Thanks to City of Watsonville staff and admin: Tami Stolzenthaler, Nancy Lockwood, Bob Geyer, David Koch, Steve Palmisano, Carlos Palacios, Carol Thomas, Clara Cawaling, and all the many others. And to council members Oscar Rios, Manuel Bersamin, Daniel Dodge, Lowell Hurst, Nancy Bilicich, and Eduardo Montesino for keeping the fire alive. Thanks to other CSW directors: Dan Sudran, Rich Bolecek, Manuel Hernandez (here it is, after all these years!), José Sanchez, George Castro, as well as all their staff—for ongoing brilliance and inspiration. Thanks to my mentors past and present: John King, Phillip and Phyllis Morrison, Paul Doherty, Maurice Bazin, Modesto Tamez, and Dan Sudran—for knowing how to pass it on in just the right manner. Thanks to colleagues who offered editorial suggestions: Sherry Hsi, Bronwyn Bevan, and Frances Gabrielson. Thanks to Jocelyn Garcia for helping with photos, Emilyn Green, Sol McKinney and Sarah-Jayne Reilly for their photos, and thanks especially to Sonya Rosario Padron for rescuing a bunch of photos. Thanks to my wonderful partner Pamela for helping me through another book. And of course thanks to my dear parents Frances and Richard Gabrielson, who always wholeheartedly supported my tinkering, sometimes to the tune of dozens of rolls of sticky tape per year.

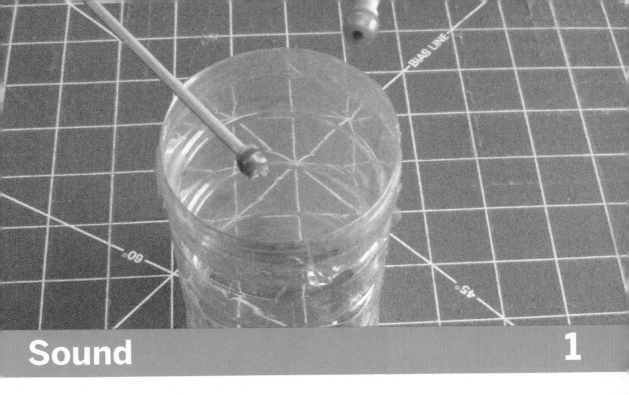

Sound   1

Tinkering with a Shrink-Wrapped Drum Set and a Torsion Drum

Sound is great to tinker with. It's rare to find a kid who doesn't enjoy making noise. Kids have at their disposal more noises than there are sections in an orchestra, but here we'll delve primarily into the percussion section. To make drums is trivial: pots, pans, cans, bottles, boxes, buckets, jugs, shells, skulls, coconuts, gourds, oil drums, and lids of all shapes and sizes become drums the moment you hit them. Put many together and you can tinker with sounds for hours with your drum set. Do it at the wrong time of night and you'll have the authorities at your door. A drum vibrates to give sound, and the pitch of the sound is determined by how the material is vibrating. Someone discovered long ago that a tightly stretched skin or other membrane can be made to sound quite nice. It turns out to be a bit tricky to stretch a skin, but with the nifty chemical technology of plastic shrink-wrap, we can all make a skin drum.

Make: Drum Set

Gather Stuff

- Stiff plastic bottles and containers (water
- bottles don't work well), cans, sturdy card-
  board tubes, and boxes, all ideally larger
  than 2 inches in diameter
- Double-sided sticky tape
- Shrink-wrap plastic, sold as *shrink film*, to
  cover and insulate windows in the winter
- Electrical tape
- Wood base and back boards (optional)
- Dowels, 1/4" or 5/16" (optional)
- Bamboo skewers
- Beads of various sizes and weights
- Craft sticks
- Other sticks to try, such as dowels, spoons,
  rulers, etc.
- Decorations

Figure 1-1. *Trying out the drum set*

Gather Tools

- Scissors
- Hacksaw or small-toothed wood saw
- Drill with bits (optional)
- Hot glue guns and glue sticks
- Hair dryer

Tinker

Basically, you'll cut out a tube for a drum body and then stretch plastic over the top of it, holding it down with double-sided sticky tape.

Step 1

Cut some of the tubes and things you gathered into drum shapes and sizes. Leave the bottoms on some of the bottles and cans, and cut some off to leave open tubes.

Step 2

Stick double-sided tape around the rim of the ones you want to attach the skin to.

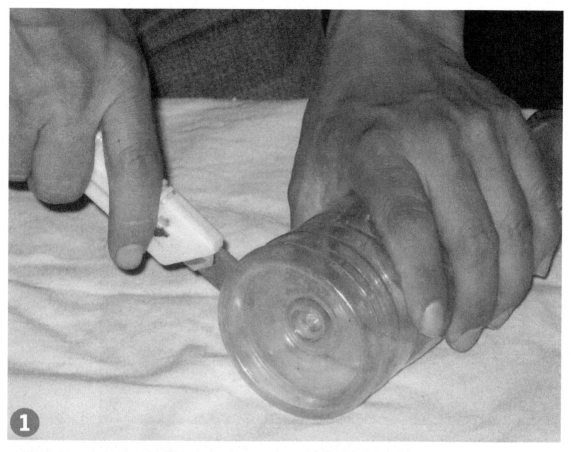

Step 3

Cut a piece of shrink-wrap plastic that is a bit larger than the tube diameter.

Step 4

Stretch it across the opening as you stick it on the double-sided tape.

Step 5

Heat the plastic with a hair dryer until it's tight.

Step 6

Trim the plastic around the tape. If you doubt the strength of your double-sided tape, go over it again with black tape.

Step 7

Try hitting it with various things, including craft sticks and bamboo skewers, both naked and with different kinds of beads on the ends. Try hitting it in the center and at the edges.

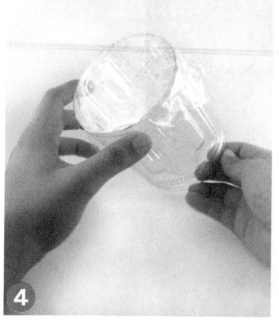

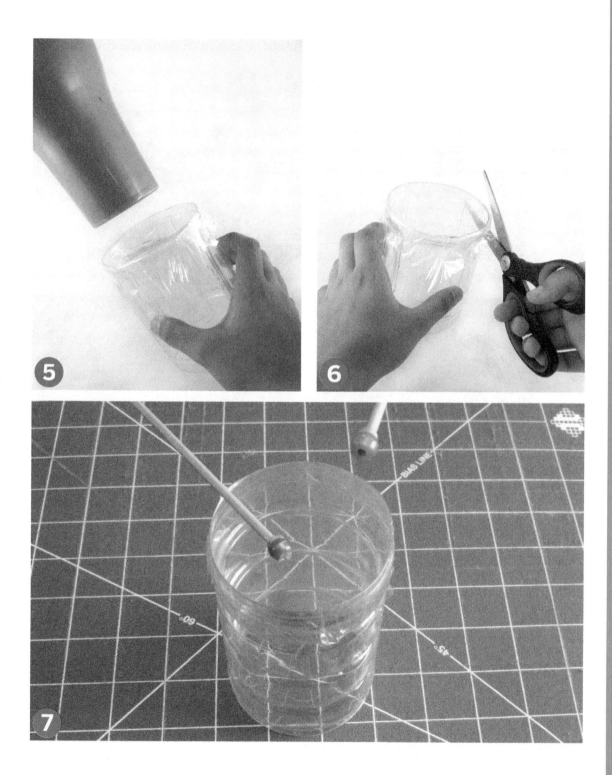

Mount your drums together into a set if you want, and add other nonskin drums and cymbals.

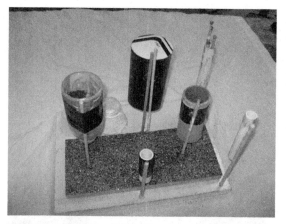 

 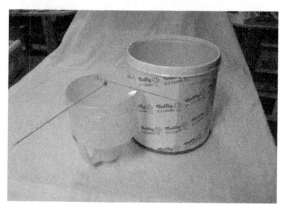

Check it out

Compare the sound of the closed drums to the ones with open bottoms. Do the drums that are sealed off sound better or worse?

- You can try drilling a hole in the side or bottom of one and see if the sound changes.
- If you like the sound better when it's open, you can cut the bottom off.
- Hit the head repeatedly as you heat it up with the hair dryer. How does the sound change?
- How does the sound change as you hit it with different sticks?
- Make at least two drums that are the same diameter and use the same material, but are different lengths. How do they sound different?
- Make at least two drums that are nearly the same length and use the same material, but have different diameters. How do they sound different?
- Make at least two drums that are nearly the same diameter and length, but use different materials. How do they sound different?

Make: Torsion Drum

Now that you know how to make a drum with a stretched head, you can make this folk toy that I've seen on three continents.

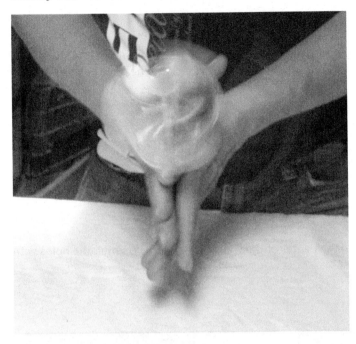

Figure 1-2. *Playing the torsion drum*

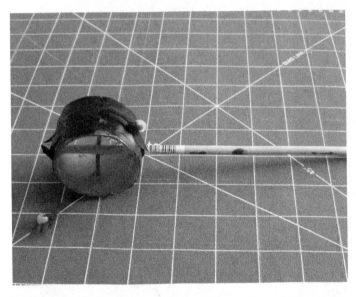

Figure 1-3. *The torsion drum in a rare moment of inactivity*

Gather More Stuff

- String
- Dowel, 1/4" or 5/16", or a new pencil

Tinker

To make the torsion drum, stretch plastic film across two heads of a short drum body, poke a stick through the center of the body, and dangle beads on strings to whack the heads when you twist it back and forth.

Step 1

Cut out the bottom of a cup or cut a section of cardboard tube about as long as its diameter. A cup is nice, since the ends will each have a different radius, giving different sounds. On the other hand, a cardboard tube can be larger and sturdier than a cup. You can use a knife to make a slit then use scissors to cut it off.

Step 2

Drill or poke four equally spaced holes in the center of the walls of the drum. We melt the holes with the hot tip of the glue gun.

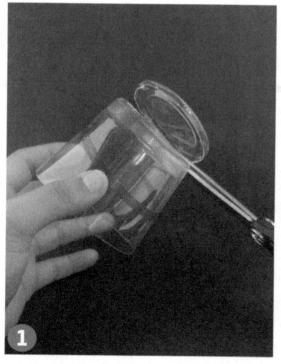

Step 3

Tie beads to the ends of two strings that are plenty long to reach from the side of the cup to the center of the head.

Step 4

Thread the strings through two holes opposite each other—beads on the outside—and tape the strings on the inside such that the beads will hit the two drum heads, one on each side.

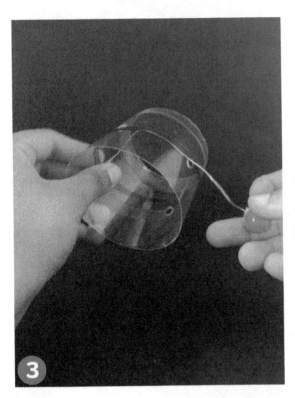

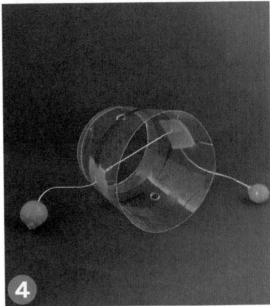

Step 5

Poke the dowel or pencil through the other two holes. Hot glue it top and bottom.

Step 6

Stick a strip of double-sided sticky tape around the edge at both ends.

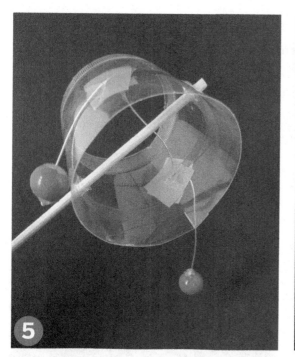

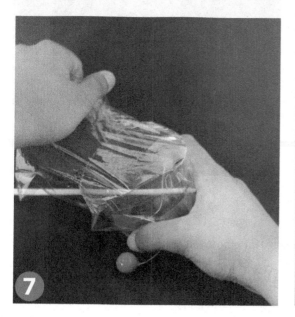

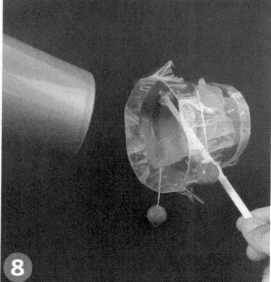

Step 7

Stretch shrink-wrap plastic heads on the two ends.

Step 8

Tighten it up with the hot hair dryer, and trim off the extra plastic.

Step 9

Twist it side to side (a movement called "torsion") such that the beads swing around to hit the heads.

Judging by your experience with the Drum Set, you can make another hole in the side of the Torsion Drum if you think the sound will be better.

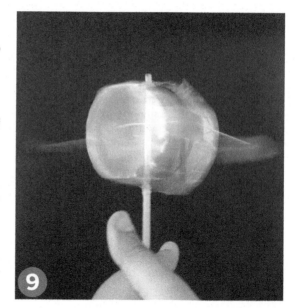

Check it out

- If you used a cup, can you hear the two different sounds made by the two different ends?
- If you used a tube with the same head size on each end, do they both sound the same? You can hold one string near the top and just let a single bead swing to hear the sounds separated.
- How fast can you go? How slow? Can you make the beads strike the torsion drum just once?

What's Going On?

To get a nice ringing sound, a vibration has to continue for a while. Any sound has one or more frequencies, and pleasant sounds often have a pleasant combination of frequencies. (A whistle, for instance, has only one primary frequency and can soon drive you mad.) Bells, chimes, and drum heads all vibrate in multiple, geometric patterns, each of which gives a certain frequency. The frequency of a sound is its pitch. A high-pitched sound means something is vibrating at a high frequency, that is, moving back and forth rapidly. This usually means it's not very large. Likewise, a large thing will generally vibrate more slowly, at a low frequency, giving low-pitched sound. Thus tubas and kettle drums give low-pitched sounds, and piccolos and violins give high-pitched sounds.

Did you find that your larger diameter drums have a lower pitch? Did the smaller end of your torsion drum have a higher pitch?

When something is stretched tightly, this usually also leads to rapid vibrations. When you heated your drum head, maybe you could hear the pitch rising. If you hit a membrane that is not stretched at all—like a sheet on a bed—there is no force to take it back to the previous position. (This force is called the *restoring force* in physics.) That's why you don't get much sound when you tap on your bed sheet and why drum heads have to be stretched.

When the head of a drum moves inward during a vibration, it is effectively pushing air into the drum. If the drum is sealed, like if you made a drum from a bottle and didn't cut the bottom off of it, then the air inside has nowhere to escape and must instead pack itself together more tightly. That's called *compression*, and it will affect the motion of the membrane. If, on the other hand, the bottom is open or if there is a hole in the side wall, the air can rush in and out as the membrane pushes it back and forth. Different drums have different arrangements for air flow. Small holes may not let the air enter and exit fast enough to make a difference.

What hole arrangement gave you the best sound?

A musical instrument always produces vibrations, and usually in the end the entire instrument is vibrating at least a bit. At the same time, there is almost always some part of the instrument that produces the original vibration. On the drum, this original vibration arises in the membrane, the head. But after a very short time, the air inside, the side walls, and even the stand holding it up are all vibrating. This is why the diameter, the length, and the material of the drum all make a difference in its sound. A drum stick hits the drum membrane, and is kicked back up by the membrane's first vibration. Thus the weight and material of the stick has an impact on the sound.

Which drum material gave you the most pleasant sound? Why do you think the best sounding stick worked better than the others?

When you compare two different drums, you can think about what is making them sound different. Different characteristics of a drum, such as its diameter, length, or material, are called *factors* or *variables*. If you want to try to determine the influence of a certain variable, you have to isolate it. To do that, you make sure all the other variables stay the same and then only change the one you're interested in. Thus, you can make two drums from the same cardboard tube, one long and one short, and then play them with the same sticks. When you compare their sounds, you can be sure that any difference arises from the lengths. (If there is no difference, you have discovered that the length variable has no influence on the pitch.)

On the contrary, if you make one drum from a small, short, plastic cup and another from a large, long, cardboard tube, they'll probably sound different, but you'll be hard pressed to tell which variables are influencing what. This is a key concept in all kinds of science; if you can pass it to your students, they'll be set for anything, and it may help them in their daily lives as well.

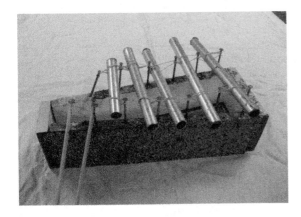

Figure 1-4. *Metal conduit xylophone*

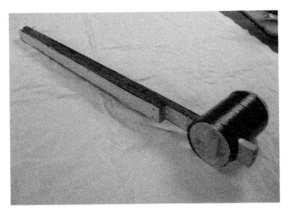

Figure 1-5. *Guitar*

Figure 1-6. *Cardboard box autoharp*

Keep On

You can make many toy and folk instruments and continue tinkering with sound. *Figure 1-4* shows a xylophone made from metal conduit tubes suspended with rubber bands over a small wood box. The box is not so important; you can also just pound nails into a board and suspend the pieces between them. We use nuts on the end of dowels to hit the instrument. The sound is surprisingly pleasant.

Figure 1-5 is a guitar we made using shrink-wrap plastic for the front of the box, 40-pound fishing line for the two strings, and screw eyes for the tuning pegs. I can actually tune this like a mandolin—the strings five steps apart—and play a song on it.

Figure 1-6 shows a sort of autoharp or zither, again with fishing line and screw eyes, this time stretched over a cardboard box, reinforced with wood. Amazing tone and volume.

Internet Connections

- Search "drum head vibration video" for many phenomenal slow-motion videos of how a stretched skin is actually moving when it rings.
- I bet you thought about how large you could go with this project. Check out images from the Drum Tower of Xi'an, China.
- Search for "vegetable instruments." Wow.

Standards Topic Links

- Sound, waves, vibrations, energy transfer, motion, velocity, and matter

More Tinkering with Music

- Dennis Waring, *Cool Cardboard Folk Instruments to Make & Play* (Sterling/ Tamos, 2000) and Great Folk Instruments to Make & Play (Sterling, 1999)
- Bart Hopkin, *Gravikords, Whirlies & Pyrophones: Experimental Musical Instruments* (Ellipsis Arts, 1996)
- Ginger Summit and Jim Widess, *Making Gourd Musical Instruments: Over 60 String, Wind & Percussion Instruments & How to Play Them* (Sterling, 2007)

2 The Value of Tinkering in the Learning Process

If you bought this book, it's likely that you already have an instinctive understanding of the value of tinkering. However, I'd like to take a few pages to expose tinkering's central role in learning and the creation of knowledge, both in the past and within today's educational institutions.

Why Tinkering Is Essential

Frank Oppenheimer, brother of Robert, who led the Manhattan Project, created the Exploratorium in San Francisco: a museum of art, science, and human perception. For over 40 years it has been an international paradigm of hands-on science museums.

Frank's great idea was that people can enjoy themselves at a deep level and learn a lot by tinkering around with things, and should thus have the chance to do so in a public space specifically set up for this activity. He knew he didn't need much funding to get started with such an endeavor, and he knew that to be a world-class museum, the public would need to be involved from the very start. Thus, in 1969, after obtaining a lease on the magnificent Palace of Fine Arts from the City of San Francisco for $1 a year, Frank moved in a lot of his old science lab demonstration apparatus, along with heaps of interesting materials, and set up a basic shop. He then put up a welcome sign, opened the door, and began tinkering with whoever came in.

Since writing this, the Exploratorium has moved to a new and bigger site across town, for which many of the exhibits have been rebuilt. Some of the originals remain, together with hundreds of fresh new ones, all allowing visitors to tinker with fundamental aspects of nature. The essence of the exhibits is a focused and particularly intimate manner of experiencing and interacting with a natural phenomenon. Many millions of visitors have been inspired by the opportunity to tinker by means of these exhibits. The Exploratorium Teacher Institute created the award-winning *Science Snackbook*, which has directions for making simplified versions of the best of these exhibits with common materials in any classroom. The fantastic response to this book enabled many more students to have the experience of tinkering with natural phenomena. Frank, who died in 1985, would have been thrilled. Back when he was a teacher, he noted that students in a given class often had no idea why they were there. He thought this was

"...a scandal. Their experience was so meager, their whole contact with the natural world so restricted, that I thought a place was needed where they could walk through a kind of woods of natural phenomena.[1]"

Later he would describe the value of the Exploratorium in terms of that personal experience:

"The notion that you can learn everything without ever doing it, as is sometimes implied in the classroom, is...outrageous, and the important thing in the Exploratorium

1.   Cole, K. C. *Something Incredibly Wonderful Happens: Frank Oppenheimer and the World He Made Up*, Houghton Mifflin Harcourt Trade, Boston 2009, p.148.

is that people feel free to touch things, to change things, to make their own discoveries... It's like the difference between teaching swimming in a classroom and teaching swimming in the bay.[2]"

But today, many students still sit listlessly in science and math class while the teacher talks about phenomena they've never experienced. Often my staff and I have seen students' first taste of tinkering to be startling and new, not like anything they've done before. Dan Sudran, founder and director of the original Community Science Workshop, once pointed out that whereas we tend to hope each kid has a bit of experience in the physical world of fixing things, taking things apart, and touching and mangling various materials, today's high-tech kids have *less* than no experience: they often have *negative* experience, in that they have really only tinkered extensively in the virtual world, where the laws of physics and biology needn't apply.

Dr. John King, my mentor professor at MIT and winner of the Oersted Prize for physics teaching, called the experience with the ins and outs of real stuff "mulch" and saw it as a foremost priority in the production of future scientists. He tinkered endlessly in his lab, which held scrap wood, metal, and plastic on basic workbenches, as well as high-tech diffusion pumps and radiation detectors. He knew full well that both low and high tech are nearly always necessary for significant advances, and so one must have a feel for both, that is, sufficient mulch from a variety of sources, if one is to truly understand and create.

Dr. King developed the X courses at MIT, with my humble assistance. These courses taught basic mechanics and electricity and magnetism concepts by means of a set of "brown bag" experiments that the students brought back to their dorms each week to build and tinker with. Through these personal experiences, they charted some of the fundamental quantities of the universe, such as the gravitational constant and the permittivity of free space.

Students in the standard, non-X courses had to accept this info on the word of authority. Students in the X courses had only to trust their apparatus. This, gentle reader, is an enormous difference. In fact, I believe it is the difference between knowledge and mere information. Dr. John King wrapped it up with his mantra:

There is no substitute for hands-on fooling around with real stuff.

When I first heard his mantra, it rang true. I grew up on a hog farm in Missouri, where I did a heck of a lot of tinkering and learned most of the valuable things in life. Certainly, you can't raise pigs on theory alone. Everything about the farming process is real. Even when you use information from some authority, such as how much bean meal to add to the corn for maximum efficiency of growth, it is with the knowledge that someone else raised a bunch of pigs with various amounts of bean meal, and found a peak on their graph. In other words, they tinkered systematically with pigs and their feed. It was absolutely *not* the case that a pig expert came up with this number by some sort of theoretical calculations behind a desk.

So it has always been natural and obvious for me that the authentic personal experience is critical. To offer such opportunities for students continues to be my first priority as an educator. I have carried out this sort of education for years and have seen it work exceptionally well in our Community Science Workshops and the sites we serve. I've seen schoolteachers pulling it off admirably in the face of stultifying restrictions in their classrooms.

I also see it at the hardware and home-improvement stores I frequent: families of all backgrounds buying stuff to *do it themselves*. They've already figured out the problem and objective; now they're out to fix it or improve it, and in the process they'll learn plenty. Doing it yourself is a branch of tinkering that is cheap, practical, fun, and hip, and it's likely to teach you

2. Frank's Quotes on the Exploratorium website: *http://www.exploratorium.edu/about/our_story/ history/frank/articles/quotes/*.

things you never dreamed of asking. *Make:* magazine, with its glorious regional Maker Faires, has been spreading the good word for years, and it's all over the Web too. Instructables.com now has posted directions for approximately a gazillion projects you can peruse and tinker from, and Maker Shed continues to offer small electronic kits on the Web and in nearly every sizable community in the nation. Clearly, this tinkering thing is not going away anytime soon.

Tinkering Throughout History

Something so phenomenally wonderful could not be new. There is great a book called "Tinkering Throughout History." It hasn't been written yet, but it's going to be a best seller. Imagine all the things we use regularly, absolutely rely on, that were developed through tinkering! The best stories will be from prehistory: fire, wheels, basketry, pottery, basic metallurgy. It will be a bit of a trick to write this book, as there will be no first-hand sources. But that doesn't have to stop us from visualizing the scene, the cave person standing beside the fresh lava flow, shielding her face from the intense heat, dipping the stick into the molten magma again and again to see it catch fire and then burn out. "This could be really useful!" she mutters to herself.

But seriously, there is a book that outlines quite a bit of what humanity has learned from tinkering. It's called *A People's History of Science; Miners, Midwives and "Low Mechanicks"* (Nation Books, 2005). Clifford D. Conner does start with prehistory and continues to the modern day, showing how knowledge has been created through people using their hands to tinker with their world. He compiles information from hundreds of sources to present a new, more accurate version of the story.

To understand tinkering throughout history, it is good to know the history of science, so here is a little primer: in the beginning, all was brutish and savage, and the hunter-gatherers knew nothing of science. Then came the Greeks, who thought up most of the useful science knowledge we know today. After that, the Romans used that knowledge to build a great empire, which soon

fell, and the world descended into darkness for several centuries while people hunted witches. Through that world of gloom and despair, a sparse few beacons of scientific thought shone forth during the Enlightenment, directing the world to ever more logic and reason and penning the science concepts we now learn in school. Henceforth, every century a few brilliant scientists operating in revered academic institutions would arrive on the scene to advance our understanding of the universe, and to them we need be ever grateful. You chuckle, perhaps, but this is pretty much what you get when you read your average account of the history of science. You are left with the impression that the greats were great solely due to their intelligence and persistence, that there was no context to their discoveries, that they figured it all out themselves by thinking hard about general principles. In short, you'd be left to think there was no tinkering, only grand experiments in professional labs, and then only to prove wise, prophetic theories, which preceded the lab work and always turned out right.

Reading the history of science is a lot like reading the history of women in a normal history book: what's missing is more important than what's written. Most history you read is just that: his story, not hers. If you want to know her story, you have to dig a bit deeper, work a little harder. Likewise, if you want to know what we've learned from tinkering, you may not get it in your average science history book.

For instance, if you are curious about where Newton got his background and motivation to work on the problem of gravity and planetary motion, you have to understand that the entire mariner community of Europe at the time was working to solve the ultra-practical problem of determining longitude at sea. Newton made use of all the empirical data and findings from navigators and ships' astronomers to form his theory. This is often skipped over by his biographers to make space for more aggrandizing praise of his persona.

Conner's thesis, roughly synopsizing, is that nearly all the scientific knowledge we have came to us by means of tinkering. Hunter-gatherers tinkered with resources at their disposal to create tools and agriculture. Ancient mariners tinkered with navigation and shipbuilding to arrive at successful methods. The Greeks, along with the Egyptians and Babylonians and Chinese before them, tinkered with what their predecessors had developed to find out even more.

Perhaps most fascinating of all, during the Renaissance and "Scientific Revolution," when the famous scientists we all know and admire made such great strides forward, there were parallel and *leading* advances being made within the craft guilds and artisan's organizations, advances we rarely read about. These vocational groups were not at the top of the social ladder, and the elite institutions looked down on them, mostly ignoring the value of their contributions. But a few broad-minded scientists from the mainstream realized that this was where real knowledge was being generated, where the real breakthroughs were happening. These scientists descended the social ladder, entered the workshops, mines, and herbariums, gathered the information available there, and used it to form new scientific frameworks for understanding.

Some of these scientists, notably Galileo and Boyle, gave full credit in their writings to the craftspeople and artisans whence they took their empirical information, yet this key element is often missing from their biographies. Other scientists tried to cover up what they'd lifted from common practitioners, claiming individual genius. Sometimes bitter disputes arose around this intellectual property. After all, the scientists were often independently wealthy, whereas the artisans relied on proprietary information to maintain an edge in the market.

Conner's examples are myriad. William Gilbert wrote a groundbreaking academic treatise on magnetism in 1600 based entirely on experimentation, something unheard of in that time. Where did he get his methods and data? Not from the scribbling of ensconced academics devoid of first-hand experience, but rather from people who knew magnets well and used them in their everyday work: blacksmiths, miners, sailors, and instrument-makers. Galileo worked out

the mathematical proof that launching a projectile at 45° will achieve maximum horizontal distance, but freely revealed that he got that fact not from theoretical prediction but from gunners at the Venetian Arsenal. Robert Boyle (of Boyle's law fame, hailed as the first real chemist distinct from his alchemist predecessors) made it clear to his contemporaries that to attain useful data it is essential to go "to such a variety of mechanick people (as distillers, druggists, smiths, turners, &c.), that a great part of his time, and perhaps all his patience, shall be spent in waiting upon tradesmen..." Conner summarizes that thus:

"the experimental method that characterizes modern science originated not in the minds of a few elite scholars in universities but in the daily practice of thousands of anonymous craftsmen who were continuously utilizing trial-and-error procedures with materials and tools in their quest to perfect their crafts.[3] *[Note: By "craftsmen," he meant craftspeople.]*"

From the dawn of time, whenever humanity has wanted to know more, we have achieved it most effectively not by removing ourselves from the world to ponder and theorize, but rather by getting our hands dirty and making careful observations of real stuff. In short, we have learned primarily by tinkering.

Fast-forwarding to the modern world, a bizarre paradox has set in. Despite a thriving do-it-yourself, Maker movement, shop classes are an endangered species, and I still know some people who don't have a pair of pliers in the house. The level of our reliance on high-tech gadgets and systems is soaring ever higher, and with it our dependency on people who can fix them, yet there continues to exist a social stigma for working with one's hands.[4] Despite a screaming need for it, tinkering is still stuck in a charming-but-unnecessary peripheral corner in society's collective mind.

To be sure, it won't be possible to go back to prehistory, when pretty much everyone knew pret-

3. Conner, Clifford D. *A People's History of Science: Miners, Midwives and "Low Mechanicks,"* Nation Books, New York, 2005, p. 282.
4. See Crawford, Mathew B, *Shop Class as Soul Craft: An Inquiry into the Value of Work*, Penguin Books, New York, 2009.

ty much everything about how to extract pretty much all the necessities of life from their environment. The mire of complicated artifacts and materials we thrash through on a daily basis is beyond the grasp of any one mind. Can you think of one mechanical object you rely on daily that you could build from scratch? I've been astonished more than once by friends who've worked 20 years in the high-tech research and development world but swear helplessness when it comes to getting my laptop back in working condition. It's easy to see how one could throw up one's hands at the complexity of it all.

Part of the message of this book is that the paradox need not stand, the complexity need not be daunting, and the stigma can be erased. It is entirely possible, for example, with utmost integrity and self-respect, to learn a good bit about your cell phone/laptop/doorbell/car/furnace/supper by tinkering around with it. The information you acquire may prove useful, or may not, but the process of getting it can be fulfilling and prepare you for further fabulous forays into learning with your hands. Those who work professionally with their hands should be lauded for their vital contributions. We should embrace the manual tinkering trades as integral to our existence and elevate these masters to the level of our academicians. This idea is not so much radical as reality—this is the way our world works.

Not for Everyone?

Some kids are not going to fall in love with hands-on tinkering. Even after tasting carefully thought out, well-presented, opportunity-rich tinkering sessions, some students will choose to learn in more abstract or removed manners. I've learned to accept this, though it took years of therapy and medication. No, but really, we all learn differently, and it makes sense that everyone's seat will not fit in any one saddle. Some people will not tinker with their hands, but rather with ideas, and ideas can also be seen as tools. (Tools that you never have to clean up and put away.) Some people will be content to embrace the computer and only tinker virtually, which is to say, not necessarily with physical reality. That said, several truths still stand.

I think it is indisputable, for instance, that all children need exposure to a hands-on, reality-based, tinkering type of learning at least as much as they need exposure to abstract and rote learning. Since the latter has come to dominate schools, I spend my energy advocating for each and every student to have an opportunity to tinker. Music offers an excellent comparison. Not everyone will be able to carry a tune, or even appreciate the different sounds of different instruments, but it is commonly agreed (though not always funded) that every kid should be exposed to a good deal of music. The same holds with using one's hands to make something: it's part of being human.

Furthermore, all sorts of things are impossible to learn *without* experiencing them in an authentic, personal manner. Take Oppenheimer's example of swimming again. Do you know anyone who says they can swim but has never jumped into the drink? How about cooking? Music? Sports? Even philosophy: I'm going to be skeptical of any philosopher who hasn't had personally moving experiences with their belief.

In fact, in addition to mere skills, a range of entire professions—some that impact humanity on a deeply significant level—require learning through hands-on, individual experience. Essentially, there are no great engineers, scientists, architects, technicians, or artists who never had a chance to tinker, play, and try all sorts of variations in their profession. For our future and their own, all kids must be given the chance to tinker.

Especially girls! I encourage you to make a conscious effort to create a space for girls. At the Watsonville Environmental Science Workshop, we have Girls Workshop once a week for an hour and a half, followed by an hour and a half of normal open door, coed workshop. We have snacks (and don't generally have snacks other times) and sometimes a special project. We put up a big banner to remind people passing by that Girls Workshop is on.

The main reason we do this is to give girls a time when the Workshop is theirs, a time when they don't have to compete with boys for space and attention. If boys come in, we tell them to come back after the girls' session is over. The boys often react with a whine, "When is Boys Workshop?" to which we respond, "We don't need a Boys Workshop."

This is reality: every day boys are encouraged by society and media to engage in activities such as taking things apart, fixing things, operating tools and machinery, designing and building things with their hands, finding solutions to mechanical or electrical or scientific problems, pursuing technical degrees, and entering technical careers. Girls are often encouraged to avoid these things or, at the very least, they are assumed to have no interest in them. Thus, while many boys feel at home in a workshop environment, girls often must go through a critical realization in which they shift their thinking to include Workshop activities within the realm of positive and doable. And fun.[5] We do everything possible to facilitate that critical realization. It often takes place together with a bunch of friends.

We put some effort into making our project models look attractive to girls (frilly decorations, such as ribbons and pompoms, and plenty of pink), but we are clear in our understanding that girls can and do enjoy *every* project that boys do, including hydraulic systems, electric cars, rockets, and even model toilets. Thus, there exist no "girl projects" in the Workshop, and we encourage girls to choose from all the projects available. We also encourage boys to learn to knit, sew, and do bead work. We also make a conscious effort to see and hear the girls in the Workshop every day, so for instance, if six kids are screaming for help, we'll take care of the girls first, with the knowledge that the boys are much more likely to return even if they get frustrated and leave.

5. As I write this, the California Makers' Network is piloting programs that bring tinkering activities to underrepresented youth, including a bunch of girls. One of the primary reasons they believe this is a good idea is that for these communities, tinkering often presents a welcoming door into science, technology, engineering, and mathematics educational and career paths.

Parents' Workbench

As parents, we're perfectly positioned to know just how our kids learn, what they know already, and most importantly, what they want to know. It's a useful exercise to sit with your kid and write a small list of the things she's interested in learning, and after she's familiar with tinkering, what sorts of things she'd like to tinker with. If you're homeschooling, this can be worked into the curriculum you've chosen to follow and can even guide the curriculum if you've got this sort of freedom. I know homeschoolers who tinker their entire curriculum.

In any event, you always want to get into a position where the kid is in the lead. The moment you end up prodding them, it is probably time to change things. Too much education is run with a bait and switch system, and you want your relationship with your kid to be on a higher level. There is a wide world of tinkering available out there, so if your kid is not inclined toward one area, have a go at another.

Don't focus too heavily on what is "supposed" to happen, and there is no need to wrap everything around a certain goal. Most important is to focus on the elements of the project or topic that your kid is interested in. Tinker beside your kids, helping when necessary, maybe working on a parallel project when everything is going well, being there for them always but sending them ahead. Think inspiration instead of instruction; once the kid is inspired, the hardest nut is cracked.

Stay flexible. Your kid may change his mind mid-project, and you may be crushed, but it may be the right move to abandon it. I once thought back on my illustrative tinkering history and figured that I abandoned over 50 percent of the projects I undertook. If you still think a given project will work, but your kid's not interested anymore, finish it yourself! Sometimes I've even been able to get my other kid to take over a project abandoned by the first kid. Kids tend to focus on the end result. We need to understand clearly that they gain mostly from the process. As a parent, you can get them to reflect on that process.

Brace yourself for your kid to wreck your tools, waste supplies, and trash your work space. It makes sense to have a separate work space for them and to keep any fancy tools you have in an off-limits place. Since I grew up wrecking my parents' tools, I have more patience but still sometimes I kick myself for letting my kids get a hold of something expensive.

Assessment with your kid can be very open and vocal. Sitting down around the glorious disaster area your kid has just generated with the most recent project and talking over what was learned, expected and unexpected, conceptual and procedural, is tremendously valuable for both you and the child. It can give direction and motivation for future explorations.

Magnetism

3

Tinkering with Magnet Toys and an Electromagnetic Dancer

Magnets scream to kids: "Play with me!" When you pass out the magnets, kids will need at least 10 or 15 minutes to muck around with the magnets before you'll be able to talk to them much or start into a predefined activity; inspired kids will need a full hour. The reason is clear: magnets give force across a distance *without contact*. Normally, this is the realm of science fiction, but here it is happening in the palm of your hand[1].

In fact, after giving many magnet lessons over the years, I've come to believe that it's sacrilegious, or at the very least criminal, to introduce the principles of magnetism without having students tinker around with some magnets first. The old adage applies with particular severity: if you *tell* someone something, you've forever robbed from them the opportunity to *discover* it for themselves.

---

1. Don't tell anyone, but gravity also does this. The earth doesn't need to touch you to pull you off the high dive and into the pool. You can also see this displayed with electrostatics: a rubbed balloon will attract your hair or a bunch of salt without a touch.

And discover they will. Every group I've ever had play with magnets that then discussed what they observed has, with only the slightest of nudge of guidance, come to the following conclusions about magnetism:

- Opposite sides attract.
- Like sides repel.
- Both sides will attract certain kinds of metal—the kind of metal that tends to rust.
- If you let a magnet move freely, and bring another magnet close by, you can control the orientation of the first.
- If you let a bunch of magnets move freely, away from other magnets, they will all align in the same orientation.
- The force a magnet gives drops off quickly with distance.

The way I do this activity is not complicated and is touched on elsewhere in this book: I let students play, suggest things to try, ask them what they're noticing, write their observations on the board, group the observations, repeat questions I hear coming from the kids, and continue to ask clarifying questions until they all agree.

At the risk of beating a dead horse, let me reiterate that students (and teachers) I've worked with, some as young as second grade, discovered and agreed on the above listed truths *without input*—not from me, not from a textbook—beyond their own observations. Of course, you can also find this information detailed in most every middle school physical science textbook, equally true, but without proof. In other words, you'll have to decide if you believe it or not, and even if you think the source is trustworthy, you won't have a feel for what's really happening. Clearly this is of limited value.

If the concept in question were nuclear reactions or something similarly complex, costly, or dangerous—perhaps we'd need to trust authority for now. But small ceramic magnets are safe[2] and dirt cheap: $0.10 a pop at the time I'm writing this. There's no excuse for not passing out a few

of them to every kid in your group. Then, after your discussion, or even before the discussion, have them tinker with making these toys and others. This activity basically entails gluing magnets onto craft sticks and onto small wood pieces into interesting configurations and exploring the way magnets work.

---

2.   Neodymium magnets are 40 times stronger than the black ceramic ones. They are available both new and scrounged from old hard disk drives. These are also great fun and instructive, but they can be dangerous. They'll smack together on your skin and give you a blood blister or break leaving a wicked-sharp edge.

Make: Magnet Toys

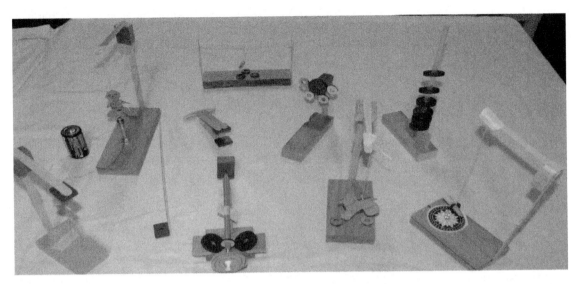

Figure 3-1. *An assortment of toys you can make with magnets*

Gather Stuff

- Magnets—nearly any size will work; here are the basic sizes shown in this chapter's photos:
  - Large ring (donut): 1.25" outer diameter, 0.375" hole diameter, 0.187" thickness
  - Small ring (donut): 0.689" outer diameter, 0.296" hole diameter, 0.118" thickness
  - Rectangular: 1" x 0.75" x 0.187"
  - Small button magnets are also useful, around 0.35" diameter and 0.25" thick, but larger magnets can be broken into small ones for this purpose as well
- Base boards, 1/2" to 1" thick, 3" to 4" rectangular
- Tiny blocks to aid in gluing craft sticks at 90 degrees
- Craft sticks and/or tongue depressors
- Bamboo skewers
- Pencils
- Straws
- Paper clips
- Kite string or thread
- Masking or electrical tape
- Compass rose (the little circle of arrows pointing to the directions and degrees of a compass—see Figure 3-11) (optional)

Gather Tools

- Side cutters
- Scissors
- Hot glue guns and glue sticks Pushpin (for compass base)
- Colored construction paper or tissue paper for decorations

Tinker

Each of the following pictures and descriptions is a just an idea, using the stuff we had hanging around at the time. The supplies listed were used to construct them, but almost no item or dimension is critical, and you could probably make them using entirely different building supplies (Legos? Toothpicks?).

Make: Floating Stack

This is the old magnets-on-a pencil trick, here on a sturdy straw braced on the base board with little blocks of wood on either side. The straw gives slightly less friction than a pencil. Each magnet threaded on the straw is repelling the one next to it.

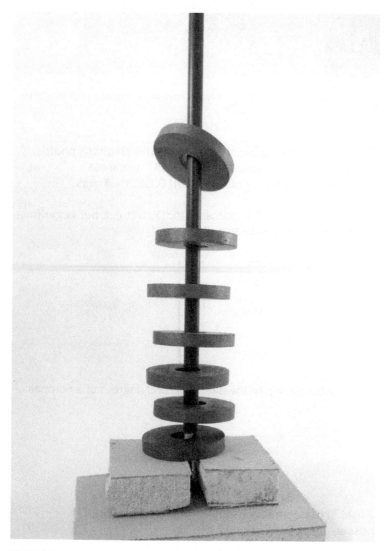

Make: Rolling Edges

Step 1

Glue one large donut magnet on the end of a stub of wood, which is glued onto a craft stick, which is in turn glued onto the base board. Then hang little donut magnets on its edge, adding them one by one. Each little magnet that sticks to the edge of the big one will repel the little one next to it.

Step 2

Glue one large donut magnet on the end of a stub of wood, which is glued onto a craft stick, which is in turn glued onto the base board. Then hang little donut magnets on its edge, adding them one by one. Each little magnet that sticks to the edge of the big one will repel the little one next to it.

Make: Hanging Up

Step 1

Make a little stand out of a base board, craft sticks, and a little wood block.

Glue one magnet up high, and let a paper clip on a string be attracted to it but not touch it. Tape off the string on the base board so that the magnet's pull beats the pull of gravity.

Make a figure out of paper—very light! Attach it to the paper clip, and it will look like it's flying.

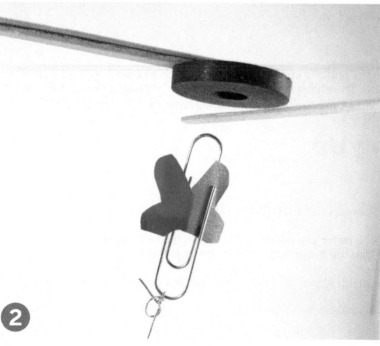

Step 2

You can also pass paper, cardboard, aluminum, steel, and anything else you want between the magnet and paper clip to see what happens.

Step 3

You can also replace the paper clip with another magnet and note the difference.

Step 4

With the right positioning, you can turn the whole thing on its side or upside down.

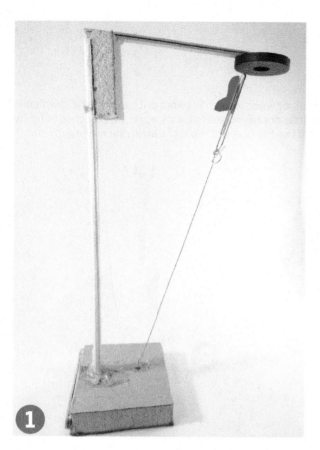

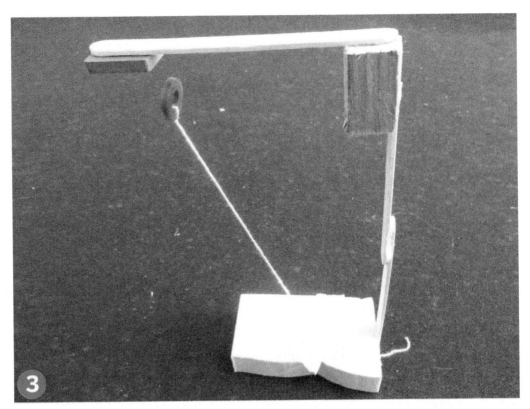

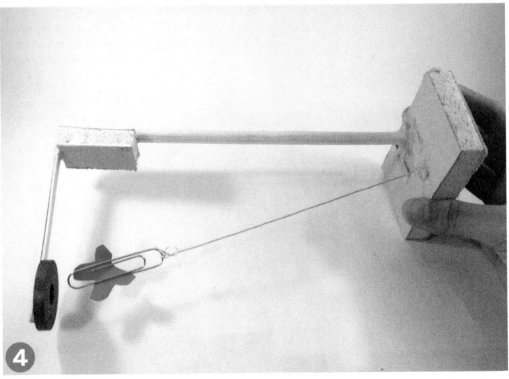

Compass

Beware, this one is mystical.

Step 1

Dangle a magnet from a single, thin thread. Now just consider it, like a Zen master. Take your time; this is good. The little, paper pointy thing through the center is optional. Try to make it point in different directions. Let it spin and stop it from spinning. Move around the room with it. Dangle other nonmagnetic things to compare their behaviors. Dangle it near metal and other magnets, and dangle it far from anything.

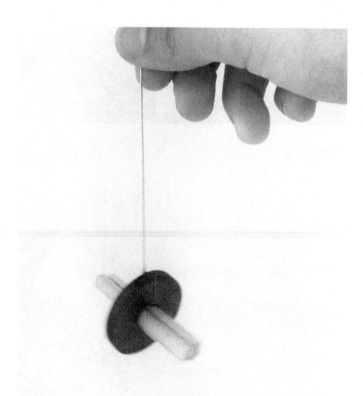

(What the heck is going on here?!) This dangling magnet, a lifeless hunk of ceramitized ferrite dust, always points the same way! It knows the directions, even if humans don't! If you are a bit taken aback, you've got good company. For hundreds of years after this was discovered, no one had the slightest idea why this worked. Indeed, many believed it was mystical, a supernatural gift. It's a compass. With it you can know your direction any time day or night, rain or shine, in completely foreign territory.

1

Step 2

Most small, ceramic magnets have poles on the faces (as opposed to the edges). Put a stick or straw or some sort of pointer through the central hole, as in the photo, mark the side pointing north,[3] and you'll have a full-fledged compass. If you forget to label the side pointing north, you're sunk; see "What's going on?" later in this chapter.

Step 3

Hang the magnet with the same sort of arrangement described in the previous project idea. Cut out a compass rose—the model is at the end of this project write-up—and stick it directly under the magnet with a pushpin right through the center. Make a mark exactly in the front. Now your compass is more functional. You can dial in a bearing, say 270°, line up the dangling magnet with north on the rose, and hike off at 270° like a professional orienteer.

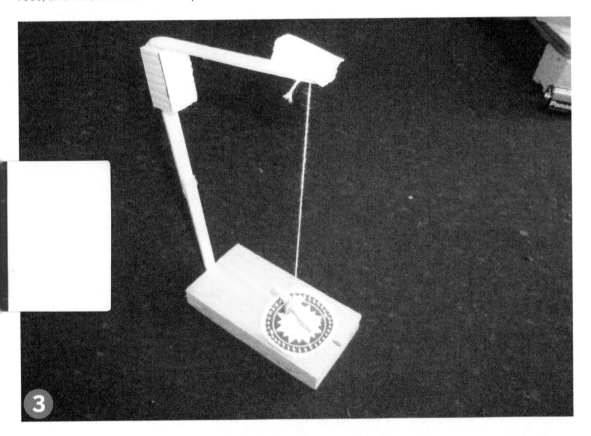

3. If you are in China, mark the side pointing south, because in China a compass is called a "point south needle," and since the compass phenomena was first discovered in China, they ought to know. (And by the way, once you know one direction, you can figure out the other three. Never Eat Stale Waffles, so they say, as they turn to the right.)

Make: Strange Pendulum

Step 1

Again with a simple stand, dangle a magnet over one or more other magnets glued down to the base board. Let the dangling one swing, and you'll see strange motion. This is an example of complexity or chaos theory. Very small changes in the position in which you release the swinging magnet will result in completely different patterns of motion.

Step 2

It's also good fun to play with if you put a little paper figure on the magnet. There are many ways to dangle a magnet, each with something interesting in store for you.

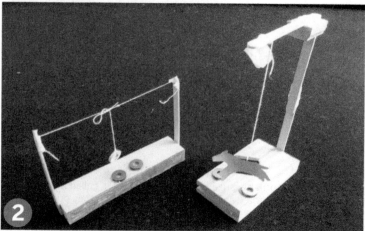

Make: Bouncing Jaws

If you press two magnets together, with like sides facing each other, they'll push apart. Let one free and it will go shooting away. Hook them together with craft sticks in a sort of hinge configuration and you'll have a bouncing alligator jaw. Leering eyes are optional.

Figure 3-2. *The bouncing jaws*

Step 1

We used tape on two sides of two parallel craft sticks as the hinge. We tried smaller hinges, but the magnets would slip off to the sides.

Make: Levitator

Figure 3-3. *The levitator*

Step 1

This is a tricky one. First build the base, more or less like the photo below. The little blocks are about 1/4" thick, and about 1 1/2" apart. Glue a long stick or dowel in a line straight out from the gap between the two similar little blocks. Glue a little block on the end of the craft sticks sticking up.

1

Step 2

Put down two large magnets, round or rectangular, about an inch apart, straight across from each other, tilted in a bit on the small bits of wood. *The faces pointing up must be the same pole*—use another magnet to test them.

Step 3

Now comes the subtle part. Thread a third, ideally smaller, magnet onto a pencil or dowel and move it to exactly the position shown in relation to the other magnets. Hold the point of the pencil against the upright block so that the pencil is more or less horizontal. You'll need to adjust the magnet back and forth on the pencil to find the right spot. You may need tape around the pencil to make the magnet stay in position. If it doesn't seem to be working, *flip the magnet over and thread it on the other way*. Only one orientation is the right one!

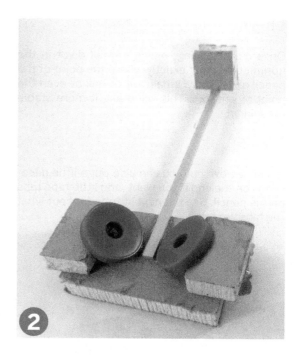

Step 4

Once it's balancing, make a small divot in the upright block just exactly where the point of the pencil rests with a small bolt or nail or even the point of scissors. This will make it more stable for spinning.

Step 5

It's done now, but you can also put a little decoration on the end if you want, and little tape tabs on the pencil. See the next step to find out what the tabs are for.

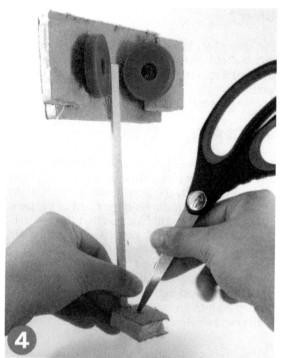

Try it out

Blow on the tabs with the straw; it should spin around while continuing to float.

Magnetic Wand (for use with all of above)

The magnetic wand—a magnet on a bamboo skewer—will have interesting effects on each device that you build.

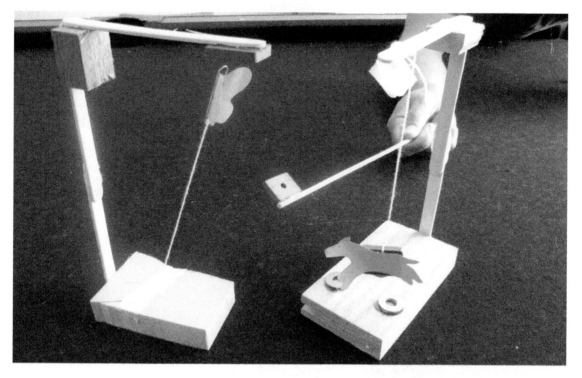

Figure 3-4. *The wand*

Check it out

- In the Hanging Up project, which materials that you inserted messed up the pull of the magnet?
- Also in Hanging Up, if you tried the magnet versus the paper clip, which seemed to pull from a farther distance?
- Why do you think the magnets at the bottom of the Floating Stack are closer together than those at the top?
- What happens when your compass is held close to a piece of iron?
- What happens if you put more weight on top of your Bouncing Jaws?
- The Levitator is only touching at one point. Can you rig it up so that there is another magnet pair at that point, so it is completely levitating without touching anything?
- What is the heaviest thing you can pick up with your Magnetic Wand?

Make: Electromagnetic Dancer

Figure 3-5. *The dancer*

Gather More Stuff

- Larger base
- Bolt or large nail
- Magnet wire—#20 to #30 gauge—or any thin, insulated wire
- Sandpaper
- Battery, any size
- Aluminum foil

Tinker

Step 1

Wrap the wire around the bolt or nail. Try for around 50 wraps, though the more you can make, the better. Leave both ends of the wire sticking out around a foot.

Step 2

Sand the tips to get the insulation off. Pulling it through a pinched bit of sandpaper works well. If the wire has vinyl insulation, strip the insulation with scissors or wire strippers.

Step 3

This is your electromagnet. You could check it by connecting the two ends onto the two ends of the battery and seeing if you can pick up some paper clips. Or you can assume it works, since it is just about the simplest circuit you'll ever make—a single wire—and glue it onto the base board. Build a high dangling stand for the dancer to swing from, directly above the magnet.

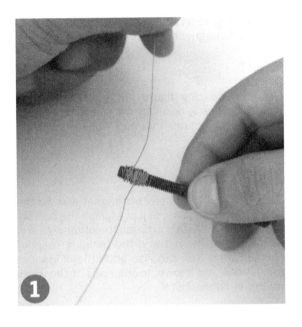

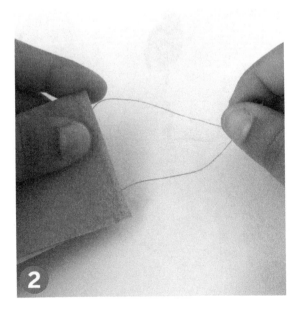

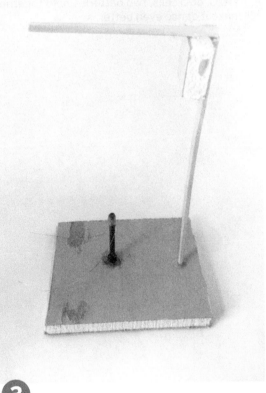

Step 4

Hang a magnet on a string or thread, as close to the electromagnet as possible.

Step 5

Cut your dancing figure out of the paper. It should not hang down below the magnet—the dangling magnet must be very close to the electromagnet.

Step 6

Tape one end of the electromagnet wire to the bottom of the battery, or just squash the battery atop the wire. When you touch the other wire end to the top of the battery, the hanging magnet should twitch. You can also add little aluminum foil mittens, not shown, to the ends of the wires for better connections.

Tappity-tap the end of the wire on the top of the battery to make the figure dance! If you're using A, AAA, C, or D cells, two batteries held together will make it dance even better.

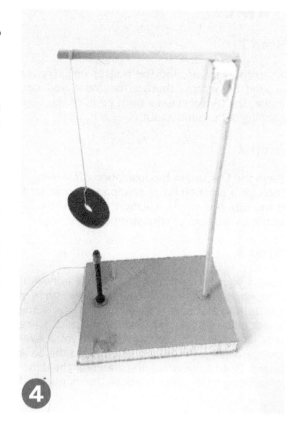

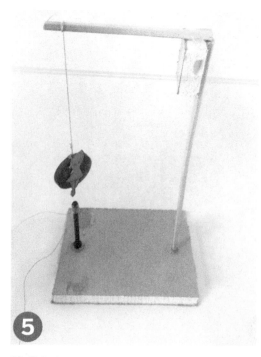

Check it out

- Is the electromagnet stronger than the permanent ceramic magnets? Test it by picking up paper clips or other stuff.
- Does it matter which way you hook up the electromagnet?
- Does it matter which side or edge of the dangling magnet is facing down?
- Could two electromagnets make the dancer perform better? If so, should their poles be aligned, or opposite?
- Can you make the dancer do the same moves with the Magnetic Wand from the previous section?
- What do you think the advantages are of an electromagnet over a permanent one?
- What do you think the limit is on how strong you can make an electromagnet?

What's Going On?

Electricity and magnetism are aspects of the same force. This force is one of the four known forces in the universe, the other three being gravity, strong nuclear, and weak nuclear. As with these other forces, the electromagnetic force can be described as setting up a *force field*. Magnets put forces on things that draw close to them. Actually, the field goes out forever, but its strength drops off quickly.

Most of these little gadgets are cool because the magnets are pushing or pulling off one another at a distance. To feel this force and see how it curves and diminishes with distance is really critical: all the equations in the world won't substitute for the real experience.

Did you notice that the hanging magnets are most sensitive to the influence of the Magnetic Wand? It's because they can turn and twitch with little friction. If a magnet is sitting on the table, it takes quite a force to make it move; to be precise, a force larger than the friction it has with the table.

Magnets have two distinct sides (or poles) called north and south. You can't tell by looking which is which. The only way to tell is to bring a magnet close to another one on which the poles are known. Then, as you noticed in your tinkering, opposites attract and likes repel. Repulsion is the special one: when you see it happening, you

have two like poles. Attraction is not so special: it also happens between magnets and things made of iron, cobalt, and nickel (this is called *magnetic induction*). So when something is attracting, you can't draw any conclusions about which pole is which.

Electromagnets work like permanent ones with the significant advantage that you can turn them off and on. A lot of work gets done this way: inside every electric motor are strong electromagnets pushing and pulling on other magnets to turn the shaft. The strength of an electromagnet depends mostly on how much current is going through the wire, how many times the wire is wrapped around the core, and how close the wires are packed to each other and what the core is made of. You can change any of these factors in this experiment and see if you can notice the difference.

Did your electromagnet act just like the permanent magnets? Did you try to find a way to turn off one of your permanent magnets?

If you stick a magnet to a hunk of iron, its field is diminished, but it's not extinguished. You may have noticed in *Hanging Up* that if you put iron, such as scissors or a metal ruler, between the paper clip and the magnet, the paper clip falls. This is because the iron messed up the field. Paper, plastic, and a lot of other things have no effect on the field. You can't turn off a permanent magnet without demagnetizing it, which means heating it red-hot, crushing it, or putting it in a powerful external magnetic field.

Compasses are simply magnets that are free to move. Hundreds of years ago in China, someone realized that the naturally occurring magnetic rocks would always turn and point the same way if free to move. So for hundreds of years, people were able to make use of that little nugget—quite valuable when you're lost on a cloudy day—before anyone knew what was going on. At some point around the 1600s, people realized that the earth is a giant magnet, and the little compass magnet is doing nothing more than lining up with the earth magnet. The very next instant they realized that they had a conundrum on their hands: The north pole of the little magnet had been named north because it pointed north, but since opposites attract, this meant that the magnetic pole near the earth's north geographic pole must actually be a *south* magnetic pole. And they couldn't just change the names on every magnet and compass that had ever been made, so they just left it like that. Ah, well. You win some, you lose some.

The Levitator makes it seem like something is floating, but the tiny place where the pencil point touches the upright on the far end is important. You can't replace that with another set of magnets. Working with just permanent magnets, not spinning, not hooked up to a computer feedback system, you can't get stable levitation. You needn't trust me on this: go try it yourself! You may find, as I did, that it's a lot like balancing one marble on top of another.

Keep On

If you make an electromagnet with a hole in the middle, that is, no core at all, you can suck iron or steel objects in. This is called a *solenoid*, and you can make it as large as you want. *Figure 3-6* is a slick little version with a straw.

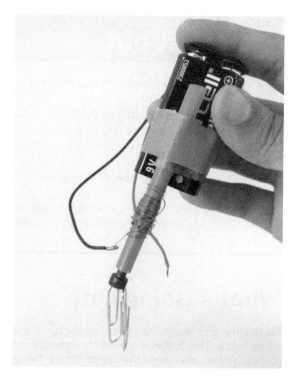

Figure 3-6. *Here you can see it's holding up the paper clip, with one unbent end stuck up inside*

Poke a nail with the tape around it into the straw, and you're back to a normal electromagnet (*Figure 3-7*).

When I was a kid I had an Erector Set, which included an electromagnet. I set up mine with a string behind the couch and a bidirectional pulley system to pick up a pile of scrap metal and move it into the back of a toy truck. Hours of glee. Forgot to take a picture.

The Exploratorium has an exhibit with a couple of gallons of magnetic sand in a box with a giant magnet (*Figure 3-8*). The sand makes spikes along the lines of the magnetic field.

You may be able to collect magnetic sand by dragging a large magnet through a sandy place. If the sand near you has no magnetite, you can also scrape the black dust off of old VHS tapes and use that. First pull the tape so that it stretches and loosens the dust, and then scrape it off (*Figure 3-9*). This will take a while, so you may also want to visit a local metal shop and sweep their floor for them.

Figure 3-9. *DIY magnetic sand*

Figure 3-7. *Just an ordinary electromagnet*

You can also tinker with magnetism using small uniform bits of steel, such as staples (already stapled or not), paper clips (*Figure 3-10*), or nails.

Figure 3-8. *The Exploratorium's magnetic sand*

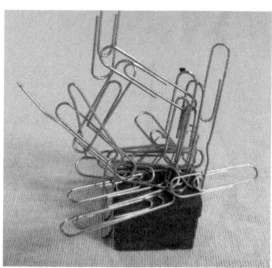

Figure 3-10. *Fun with paper clips*

Internet Connections

- Search for a video about a "levitating frog." The magnetism we commonly talk about, that only works on iron, cobalt, and nickel, is called *ferromagnetism*. It turns out there are two other types—*diamagnetism* and *paramagnetism*—and that many things can be pushed on with very strong magnets due to their diamagnetic properties. No harm done to the frog, mind you.
- Search "superconducting magnetic levitation" to see how magnets will levitate stably over a chunk of conductor that is cooled to the point that it has no electrical resistance. This is due to induced *eddy currents*, another fascinating branch of magnetism.
- Check out "drop tower rides." These amusement park rides drop you from way high in the air (100 to 400 feet) and then stop you at the bottom with magnetic eddy currents.
- Look at *http://www.levitron.com*. This one also actually does levitate, like I told you wasn't possible, due to stability gained by spinning.

Standards Topics Links

- Electricity, magnetism, circuits, compass, forces, and fields

More Tinkering with Magnetism

- John Cassidy and Paul Doherty, *Magnetic Magic: Magic Tricks Done with Magnets* (Klutz, 1994; a modern classic—out of print, but I hear a new edition is on its way)
- Fred Jeffers, *Mondo Magnets: 40 Attractive (and Repulsive) Devices and Demonstrations* (Chicago Review Press, 2007)
- Simon Field, *Gonzo Gizmos: Projects & Devices to Channel Your Inner Geek* (Chicago Review Press, 2003)

Figure 3-11. *Compass rose template*

4 A Good Tinkering Session

What a Tinkering Session Looks Like When It's Working

A lot of times good tinkering just happens. This can be a goal: to get our kids to the point that they'll feed contentedly on a hot-n-ready tinkering opportunity and even be able to cook one up for themselves. But many times, especially at the beginning, you'll need to put some legwork into setting it up. To facilitate good tinkering, you should think about the general scheme you want and the grand framework for the tinkering you'd like to see happening. There are some key characteristics you should look for in the tinkering session. It's also good to think a bit about the kids you'll be working with.

Tinkering Schemes

At the Community Science Workshops of which I've been part, we engage in more than tinkering. We're a community resource for education of the hands-on, technical sort, and we attempt to serve whatever the community's needs are in this area. That said, a whole lot of what we do is tinkering. Our tinkering can be divided roughly into two basic structural categories:

- Free-form, aka open structure, aka, at times "madhouse," "nutso," or, "Who the heck is in charge here!?"
- Class group, aka "all-on-the-same-page"

Free Form Tinkering

At the Watsonville Environmental Science Workshop, we especially value our freeform tinkering. Forgive my unbounded self-satisfaction, but I view it as a true paragon of education and learning. When we're tinkering up a storm, everyone present has chosen to be there, at that time, to work on a project of their own choice. There are no institutional restrictions, no time limits, no tests or evaluation of students, and no curriculum beyond the projects' models and materials hanging about, a small set of books, and the Internet in the corner. Daily we see dreams and ideas turn into reality as life-sustaining skills and confidence are built on a solid foundation of joy, all in a comfortable, unstructured, and supportive atmosphere. Educational paradise, I've said to myself many times, couldn't look much different.

Thanks to the good support of the City of Watsonville and our various donors, we currently offer free-form tinkering at seven sites, 15 times per week. Woohoo! Kids come in and are able to do *whatever* they want with *whichever* materials we have on hand, and we help them the best we can. We'll even get more materials with our meager budget to continue the tinkering tomorrow, look up information on how to do what the kids want to do, and call up experts if we run out of expertise. Our Workshop is set up for awesome tinkering on demand in the areas of sewing and needlework, microscopy, electricity and electronics, magnetism, music and instrument making, woodworking, metalworking, plastic forming, cardboard construction, dissection of both animals and appliances, papier-mâché, safe-and-sane chemistry, bicycle maintenance, simple optics, arts and crafts, astronomy, gardening and potted plants, cooking, aquarium life, small animal behavior, and probably others.

Kids come in whenever the door is open. It's a bit like a recreation center, only with many opportunities for full-content, full-throttle learning. Most kids wander around a bit and see what's new and then choose an area to sink their teeth into. Others wander all afternoon. Still others ask us what they should do. We'll help them find their ideal project and others to work together with, but we'll never coerce them to do anything. If they insist on causing trouble, we'll kick them out (into the park outside), but only for a while; we'll even follow them home to be sure their parents know we want them back in the Workshop when then can follow the rules.

Speaking of rules, we've got only a few solid ones, aside from the safety rules described in the Logistics section. Here they are:

- No food in the Workshop, unless I get some.

Just kidding, but really, I've gotten through many a long, hungry afternoon with that line.

- Respect all (stuff, one another, yourself).
- Clean up after yourself (ha!).
- Get smarter!

The first one pretty much covers everything. Like the golden rule, if everyone would but follow it, things would be dandy. For example, we could have a rule that says, explicitly, "Don't screw up other kids' projects," but the first rule rules in this case.

Try as we may, we can't seem to enforce that second one, but we still have hope, and it gives our lives meaning to think that once in a while maybe one kid will learn a few cleaning-up skills and surprise his mom when he gets home.

The third rule is a guard against lazy butts just sitting on their cans in the midst of all our fruitful activity. They're a bad influence, and we've found we can't let them crop up and multiply, so we'll tell them they have to tinker with something or they're not getting smarter and they'll have to leave. Sometimes it works. Of course, we reserve the right to tack addendums on to any of these rules anytime we want.

Sounds pretty wonderful, eh? Almost surreal. Before you write me off as a teller of tall tales, let me make clear that it took us many years to get the space set up for this. It's quite a subtle challenge to have everything available for all these areas of tinkering and still keep chaos at bay. Not to say that we are always successful at keeping chaos at bay, but often we are. We often get 30 to 50 kids in the main Workshop all jamming on great tinkering projects. We have two adult staff present at all times and try to have at least a couple of high-school helpers on duty as well. When a day is really hopping, we'll be bouncing around like crazy from project to project, and inevitably some kids may get frustrated and leave. But they can return another calmer day and resume where they left off.

Project models are key to our operations. We try to maintain a good 50 or so project models that span the areas previously listed, all hanging invitingly from the walls and ceiling. It is a challenge to maintain them while also letting the kids take them down and use them as a model to work from. But it's worth it because models allow kids to tinker toward a goal in a way written directions never can. Models also allow for a bunch of kids to work on different projects without step-by-step assistance. Learning to follow the model instead of cookbook directions is another element nearly extinct in our schools.

Not everything needs a model. For example, the under-table storage bins full of messed-up household appliances are a no-brainer. Particularly for your wild, freespirited kids, you may just need to encourage a bit of analytical destruction and they're on their way. We also have project write-ups for many model-less projects in a little folder with photos and all, but this is most often used when a kid has exhausted interest in our set of models. Many of these write-ups have cookbook directions, but even these directions offer various paths to various goals. We attempt to avoid cookie-cutter projects at all costs.

We have this free-form scheme running at our satellite sites as well, where we drive a van full of tools and materials and project models to locations with high kid concentrations. We unload our stuff—a few boxes of junk, a set of plastic drawers full of supplies, and a desktop scroll saw and hand power drill—and haul it all into the community space, whether it be a park or an apartment complex, then let the tinkering begin.

We also do limited free-form tinkering at events such as Earth Day or the Strawberry Festival, or at a family science night at a school. We'll have piles of just a few different materials toward a couple of project possibilities, and let kids and families tinker with them in whatever manner they so choose.

Vignette: Tinkering with Gak

Gak, described briefly in Chapter 9, can be made with specific cookbook directions. Each kid can end up with a blob identical in texture and size. We prefer to tinker with our gak. When someone wants to make it, we get out the box of supplies and let each kid decide how much of each ingredient to put in. The variety of results are then compared and discussed. Tomorrow we can do it again. Some kids will do it every day for a week, each time learning a little bit more about the way substances interact.

Class Group Tinkering

The class group scheme is what we use when we go to schools or when we receive set groups of kids at the Workshop. This may take place either during the school day or after school. We take materials for a single project and let everyone tinker at making their own, either alone or in groups. We prepare one or two working models of the project and show how it works at the beginning of the period. We show more or less how to tinker it together, giving detailed directions at tricky or dangerous parts and vague directions for the rest, leaving plenty of room for creativity and new ideas. Then we ask a couple of focus questions to get them thinking about how it works and spend the rest of the period tinkering. Five or ten minutes before the end of the period, with luck after a bit of cleanup has occurred, we attempt to sit the students down in a tight group and have a discussion about what happened, what they observed, and what they learned. We'll go back and address the focus questions if anyone is interested.

In this scheme, we're limited by the time restrictions of an institution and also by the fact that on any given day, some students are there not by their own choice. This difference is vast and significant. Tinkering doesn't necessarily fit well into time restrictions. When kids know they may not be back, we can't encourage large projects.

When everyone is tinkering toward a similar goal, there are inevitable bottlenecks, at the drill station for example. And finally, just one kid who's being forced to participate in something she's uninterested in can sour the whole room.

Still, there are advantages. Over the weeks, we can run through a set of projects that give kids a deeper view into a topic, such as electricity and magnetism or light and color. We'll choose projects we know to be widely popular and encourage uninterested students to alter them to their own satisfaction, or even tinker with the materials we've provided toward a completely different project vision.

It's nice to have everyone on the same page when we're discussing what's going on; everyone can offer observations from their own experience. It's nice to have a final discussion—closure is good for everyone—though in our experience, it's the single most difficult part of the day: the minds of kids who've been taught at all day long and are ready to go home are not necessarily open for more pondering. Sometimes we try to do the closure in small groups as they're finishing up their tinkering. This way there is less tension and stress when kids are ignoring the discussion and goofing off.

We do our class group tinkering with two high-school helpers whenever possible. The more help you have, the less direction you have to give, and the more you can let kids figure things out themselves. The limit is often the kids' toleration of frustration; we want to give them the valuable chance to struggle through a problem without assistance but also want to be able to run over at the last minute before they chuck the pieces in the trash and give up for life. Clearly some students are more used to this model of learning than others. More on these issues shortly.

Vignette: Dan's Oscilloscopes

Dan Sudran, grandfather of the Science Workshop, saw that kids have trouble conceptualizing the difference between AC and DC electricity. He put together a bunch of simple, low-voltage AC power supplies so kids can hook up different components and compare the result to batteries, which are DC. He also scrounged some old oscilloscopes, instruments that allow you to see the otherwise invisible electrical signal. Oscilloscopes are quite complicated, but Dan presets the controls so young kids can just see the signal from their circuits. He sets up stations around the room, distributes the students among the stations, and instructs them to tinker. Then he and his staff wander around tweaking controls, appreciating discoveries, and encouraging investigation. Kids lead, facilitators back them up, and discoveries are discussed. What a difference from the lock-step, brain-numbing, soul-crushing lab exercises common in most Intro to Electricity courses!

One final note on tinkering structures: at the Watsonville Environmental Science Workshop, we rarely carry out the standard laboratory exercise, with set procedure, single phenomenon analysis, data taking, and report. We rarely do an "inquiry exercise," where the students are walking together with the instructor through a limited set of experiences, talking about them as they go. We also lecture to the students as little as possible. To be clear, we don't think any of these are ineffective or unwise ways of doing education. We do see that schools have more or less got them covered. Many schools are extremely constrained in what they can do, and while we may try to influence them to change and broaden their vision, at the Workshop we strive in large part to do what schools don't. Thus we try to maximize freedom and student choice, maximize the possibilities of pursuing personal fascination, maximize opportunities for students to seek their inner genius, and maximize joy.[1]

Frameworks for Tinkering

Aside from choosing the structure in which the tinkering occurs, you'll need to think about the general framework you want to create around it. The results will be quite different within different frameworks. I'll describe four here, and certainly there are others.

Studio

This is our modus operandi for the Watsonville Environmental Science Workshop: free-form, open door, open structure—a big welcome to all to start and continue tinkering projects of their

---

1. As my mentor-professor John King used to say, if schools were doing this already, and only this, we'd be busy organizing study groups to work on regimented book learning and problem sets Clearly one needs both to learn well. It's an enigma as to why the school pendulum has swung so far in the direction of rote learning. Some say the pendulum is off its hanger.

choice. We provide a sort of community space for tinkering. I've likened our Community Science Workshops to libraries, rec centers, and clubhouses, and here's another comparison: a communal artist studio, where the materials and tools are shared and the goals are determined by the participants. Our place is really well set up for this, but I've seen classrooms transformed into very nice tinkering studios, too.

Competition

This is a common one in school science class or engineering clubs: make a device that beats the others. Good fun and the quintessential external motivation,[2] a competition focuses the tinkering toward a singular goal. Most engineers end up competing regularly in the reality of the business world, though environmental engineers often have broader goals, and public works engineers must cooperate with many and varied interests. Don't forget that in nearly every competition the majority of the kids lose. This is not insignificant.

Cooperation

This is also popular: "Let's all work together to build a duct tape boat!" The bigger the project, the easier it is for a lot of students to be involved in it. But kids all tinkering on separate projects can also be cooperative because they're trading ideas, giving suggestions and critique, etc. The cooperative tinkering can also be toward a reallife issue, such as growing gardens or understanding bicycles enough to fix them. (Of course, you can encourage cooperation regardless of the framework. I think it's always a good idea. The space can be arranged so that it's easy for students to work together. The facilitator can set the scene, perhaps with a, "Hey, be sure and check out one another's projects; you may get some good ideas and may even want to work together.")

Individual Exposé

This is the science fair model, often quite competitive, but unnecessarily so. Instead of giving medals to the "winners" while declaring that winning isn't everything, why not a certificate to each entry describing the unique excellence that it has? You needn't coordinate with the local branch of the national science fair if it seems too difficult or limiting. Each kid or group of kids can work on a tinkering project with the goal of showing it at a public event or to a group of people (bigwigs, parents, younger kids, peers, and old folks). I've done several "mini-museums" this way, and Mini Maker Faires following this model are sprouting up all over. An external audience often changes the way things go and is a good experience for kids. Having to explain and answer questions about your tinkering is fine brain exercise.

2.   See Alfie Kohn's *No Contest: The Case Against Competition* Houghton Mifflin, Boston, 1992, 2nd edition, for an excellent exposé on how little is gained and how much is lost by emphasizing the fight.

Vignette: "Required" High-School Projects

Each year we get groups of high-school students showing up at the door with plans to do a science project. A required science project. The goal? A decent grade. The way they drag their tails in makes one's heart sigh, but we're always encouraging because we've seen enough examples of what you may call nontech kids getting honestly wrapped up in the project to the point of busting through one or more frustrating challenges to emerge victorious with a working wind generator or musical instrument that can play an octave scale or whatever. Heartfelt thanks are common and even hugs are not unheard of as they walk off cradling their precious creations.

Characteristics of a Good Tinkering Session

Facilitating tinkering well is much like excelling at any art form. It entails nailing a whole range of factors spot-on every time. Here's a list of some of those factors in no particular order.

The session is focused on the students

Not the final projects, not the materials, not the tools (valuable as they may be), not the concepts you hope the students learn, and certainly not you. You are accompanying these youngsters in the process of learning to tinker and learning from tinkering, so focus on them.

The facilitator is not lecturing much to the group

I'd say 20 percent of the time is an absolute maximum, and that's only when you are dealing with a complex topic *and* the students are way interested in it. We routinely talk for less than 15 minutes out of a two-hour class group session. At the free-form Workshop, we may not talk much at all beyond giving occasional, on-demand pointers. We've noticed that this can be tough for experienced teachers; old habits die hard. But the point here is that the student will learn primarily from the stuff, only secondarily from you.

The facilitator is engaged nearly constantly with the students

This one is a particular challenge for me. When everyone is tooling away on their tinkering, sometimes I like to just let them go. It's OK to just tinker quietly along beside the students, but it is nearly always beneficial to get a student to talk about the current state of their project. If one doesn't want to talk, others will. Aside from developing language and communication skills, they're crystallizing what they're doing, planning, and learning.

Students are working together

Students often want to build their own project and often *don't* want to help one another, but we still encourage them to work together as much as possible. They'll learn more that way, even if one of the things they learn is that they'd rather tinker solo.

Many materials are available, and many options are open

We try to approximate the universe in our Workshop, and it's not unreasonable. In the end, we all live in the universe, so what argument can you make for the benefit of limiting this reality for your students? Essentially, all limitations we do impose are natural—no, we can't build a nuclear reactor, no we can't experiment with real gold—as opposed to arbitrarily determined by us. As I learned from my friends who work in early child development, we look for every opportunity to say, "Yes!"

Questions are thick in the air

How does this work? Have you ever seen anything like this before? Why is yours running so much smoother than mine? Did you see my cousin's version of this, with the car battery? Do you remember learning about this in school? Is this like what we did last year with the egg cartons and syringes? Is there any way to make this work better? How can I change mine into a birthday present for my mom? How come when I do this, that happens? What the heck is going on here?

(This one's a bit nebulous...) There is an atmosphere of joyous desperation

You've seen kids offered some challenge, such as getting a ball out of a tree, and go nearly berserk falling over themselves to solve the problem. That's what we're looking for. That's the mind wide open and ready to suck in any new knowledge or wisdom that floats by. Of course, there is rarely any true time pressure for anything when you're a kid, but when you're keen to work on something, you want to really go at it! Alternatively, you can also aim for the calm, steady, Zen sort of tinkering that you'll often find in violin makers or jewelry crafters. You may have a different population of kids than me—I think I've achieved that with my students two or three times in the last 20 years.

Clear order exists amidst the noise and mess

Again, it's a balance, but entirely possible. For safety's sake, the floor must be reasonably clean and the hazardous tools must be used in an orderly fashion. It must be clear who's in charge and the boundaries of behavior must be well known by all. Beyond that, take an aspirin for your head and revel in the good, good learning.

Students

(Here I'll be talking in broad generalizations; obviously each kid is a special gem of unfathomable potential, always changing and growing. Don't fall into the treacherous trap of thinking that the one who's the bane of your existence will always be, *or* that the little genius is the one heading for greatness. Still, I think it's useful to lay out these general thoughts.)

At the Workshop, we see some kids who've never touched a tool, never been given the opportunity to solve a technical problem, never learned to tie their shoes (thanks a lot, Velcro!), and who turn to authority to know what to do in any given situation. It's an irony that often this kid resists listening and following directions, since in the end, that's all he really knows. When away from authority, he'll often follow his buddies, which is not an entirely healthy way to live. A good tinkering session can be a life-changing experience for him.

With this kid in the free-form tinkering scheme, we've found that we'll need to suggest a project, put him with others already tinkering with a goal, or stay with him until we've found a good match and he's well plugged in. This kid can be pretty fragile. Once in a while, we've been too busy with other kids to stay with him, and he leaves frustrated, never to return. In the class group situation, he'll often be the screwball, disrupting to avoid showing his lack of comfort, claiming boredom (yeah right, with all these tools and materials?), or just shutting down. It takes insistent, calm, repetitive invitation and just the right amount of assistance to make it work for him. Your first goal as facilitator is for him to have a good taste of success.

Swampscott Public Library
61 Burrill Street
Swampscott, MA 01907

Vignette: Javier's Bike Tire

Javier feared tinkering. He feared failing, and had been labeled a failure in school for years. "I'm not good at building things," he'd say, rejecting our invitations and heading to his daily solace: violent computer games. Everything changed when his mom bought him a bike and the inevitable maintenance become necessary. "Can you help me fix this?" he'd say, leaving it in a pile and heading for the door. When he returned we hadn't moved it but stood ready to help him fix it. He'd grab it up in frustration and walk it home. This repeated several days in a row, until he grew so desperate that he was willing to give it a try. It's not rocket science to patch a tire, but there are several key steps, and once he mastered them, we were sure to point out what he'd learned: he was good at this, and he could do it himself. It wasn't long before he was teaching his friends. "It's easy!" he'd proclaim, without irony.

We also get master tinkerers, some of them know-it-alls. Well, come right in! Sometimes she'll have a project idea all ready in the incubator that is her mind. She may ask for assistance or she may not. A lot of times she'll run into obstacles and need our help. In the event that she's not jamming away on a project, we'll put her to work helping others if she's keen to, and regularly pump her for new ideas. Some kids like this can be on the lazy side, in which case they'll build a few awesome projects and then complain that "there's nothing new here." *Oh, no, no, my friend: that's why you're here! I have this funny little motor and gear setup someone ripped from a DVD player, and I was thinking you could hook it up to this toy bear to get a reciprocating motion.* We truly do look to these young masters for new ideas, new versions of old tinkering projects, new life from old junk. We've seen time and time again that the best ideas for kids come from kids. Most kids come into the Workshop, wander around, choose something to do, and start work-

Vignette: Silvino's Rubber Band Gun

Silvino was one of our consistent sources of new projects. Once he came in talking of a rubber band machine gun he saw on YouTube. We have a firm no weapons policy, but I sanctioned the project because he was so jazzed and it seemed harmless. He built one straight away, even adding a mechanism for a Gatling gun crank trigger and detachable rubber band magazines. About 10 minutes after his completion, a group of kids came in from the nearby apartments and pounced on the project. We built 10 of them before the day was done.

ing on it. In the class groups, most kids see the project, are interested in trying it out, and start in at it. As they begin to work, you can soon get a feel for their previous experience in that given area of tinkering, their dexterity with this set of tools, and also their general ability to solve problems and figure things out. Your facilitation then should respond to their realities, here and now. Frustration is a common response to tinkering. Dealing with and overcoming frustrating situations is key to successful tinkering, and I would say also to a happy life. Here the tinkering facilitator can really make positive changes in kids, though sometimes it's painful. A key question for facilitators is this: what's your response to a kid whining to you for help when he could very well figure it out himself by looking around and asking his colleagues for direction? This situation arises nearly every day for us, and it takes some skill and fortitude to confront it gently, firmly, and effectively each time.

This frustration management is the very core of what we offer by facilitating tinkering. If kids can handle their own frustration, they won't need us to help them tinker. Many, many *do* need us desperately and are much for the better when, with our help, they've worked through a snag. Think what you may about the relative benefits to humanity of various professions, but when I've helped a whiny kid to understand that she can do it herself or that she can get help from any number of sources or she can figure out what's holding her back and fix it, I'm very much satisfied with my place in the universe.

I tell my students that when they're feeling extremely frustrated, they should also be experiencing a rising joy, since they're nearly at the point of breaking through. That "tinkerer's high," like the surge of a long-distance runner, is not only pleasant and fulfilling, but also an essential ingredient to the overall experience of a scientist, engineer, technician, or artisan. Keep your eyes open for it, both in yourself and in your students.

Vignette: My Tinkerer's High

When I was in high school, I took an electronics course and entered a contest to design a project that used batteries. It was 1986, only a few years since LCD watches had become common, and there were no mobile phones or pagers with their familiar vibration modes. I had one of these new watches, and it regularly went off and got me in trouble in church or class. I wanted to be able to get the alarm reminder, but not have it make noise. I thus designed a tiny solenoid that would poke on your arm with a blunt nail, and hooked it up to the watch's speaker through a silicon controlled rectifier. When the watch sent a current to the speaker, my solenoid would cut loose. So far so good, but instead of going poke-poke-poke like I envisioned, it only vibrated.

Essentially, the feedback circuit was too fast. I knew very little about electronics at the time, but I was also young and confident and undaunted. One evening, after several hours of unsuccessful tinkering, I got frustrated and went to bed. I lay there staring at the ceiling and thinking of all the different electronic components I'd been learning about in class: resistors, inductors, diodes, capacitors; wait, now what is the function of a capacitor? To store a charge! Wouldn't that do the trick? I ran back to the work bench and found the largest capacitor that would fit in the device and hooked it up in series. Now instead of vibrating, the capacitor took a small fraction of a second to both charge and then discharge, which slowed the circuit and provided a nice, steady poke-poke-poke. I rode that high for days: I had not only fixed my fabulous project, I had made a huge leap in understanding capacitors. All on my own, all through tinkering.

Certain students will benefit tremendously from the hands-on tinkering experience. These students have had a rough time in the classroom, with great difficulty in learning and remembering what they're "supposed" to know. They've been told repeatedly by institutions and by individuals that they're not smart and not good at various skills. Not all students that schools have labeled failures will thrive on tinkering, but a good chunk of them will. Maybe the kid can't or won't listen and follow directions, but when she's given freedom to figure things out and create things herself, she can shine. If she's the communicative sort, you'll hear about it as she walks away: "I'm good at tinkering!" or "I can make anything with tools!"

The kid's found her genius. She's on her way to self fulfillment, which means there's less chance she'll be a drag on society when she grows up. It's heartening to know that years of oppressive, nagging degradation in our homes and institutions can be countered by a few successful flings in the realm of tinkering. The positive experience you facilitate with a kid tinkering may have multitudinous ramifications in their future. They may have renewed patience to try to learn at school and new dreams of a potential career. As Dan Sudran says, we're not just messing around with kids and junk here, we're changing the world. Keep the faith.

Vignette: Victor's Delight

I overheard a rambunctious fourth grader named Victor ask his buddy if he still had the project he made last week. I forget his buddy's response, but Victor's animated reply was one of the best bits of feedback and affirmation I've ever received: "Well, *I've* still got *my* projects at home, but I took them apart and made all kinds of *other* projects out of them!"

Mechanics   5

Tinkering with a Basketball Hoop, and a Carnival Ball Game

Mechanics is the area of physics involving force, motion, and energy; in other words, just the sorts of things auto mechanics think about when they try to fix your car. Many of the things you do that are a lot of fun involve serious mechanics: roller coasters, martial arts, fast cars, sports, trampolines, pillow fights, etc. Tinkering with mechanics may be tinkering in its most elemental form. Fooling around with some cool gadget, trying to understand it, trying to control it, and make it work better: these are the flesh and blood of tinkering.

Many people shy away from physics due to perceived complexity. The irony is that many truths in physics can be stated in utterly simple form. "Uncle Isaac" Newton laid down three laws of motion that wrap up a lot of mechanics. He said things will stay in their current state—in motion or at rest—until an external force acts on them to bring them out of it. He said the acceleration of an object—how fast its speed changes—depends on its mass and the force it receives. And he said that every action (and by that he meant a push or a pull) has an equal and opposite reaction. Everything that moves follows these rules.

Now that you're up to date on the physics of mechanics as of 400 years ago, let me just say that these projects comprise some of the most popular of all the tinkering that happens in the Watsonville Environmental Science Workshop. Sending something airborne, even briefly, is what fun is made of. It can be a bit dangerous, perhaps, but since we all know that safety is first, we take all necessary precautions, and no one gets hurt. And once the force, motion, and energy get moving, there are plenty of questions to ponder.

Make: Basketball Hoop

Gather Stuff

- Longish base board, 1" or 2" thick, 3" wide
- Dowel, 5/16" or larger, 1' long or so
- Two 1" boards
- Small block of wood
- Backboard of cardboard or similar stuff
- Plastic bottle neck
- Plastic spoon
- Duct tape
- Small nails
- Thick rubber bands
- Puff ball or similar harmless ball to shoot

Gather Tools

- Saw
- Hammer
- Drill with bit that makes a hole the same size as the dowel
- Hot glue gun with glue sticks

Tinker

The parts list is for the project shown. You should use whatever materials you have sitting around to get basically this arrangement: a bottle neck or any sort of cup as a hoop with a modest court in front of it and an enormous backboard behind to save you endlessly running after the ball. You want the shooter system to be adjustable in at least two ways: the angle and the distance. Here the dowel is hammered into holes drilled in both the base board and the smaller board that the hoop is nailed to. If you can get a drill bit slightly smaller than the dowel—19/64" for the 5/16" dowel, for example—you can hammer the dowel in tightly and not even need glue.

Step 1

Make a court, hoop, and backboard. We painted our project at this step, since we didn't want the paint to foul up the shooter mechanism.

Step 2

Fasten the spoon onto the front edge of the movable wood block with a duct tape hinge.

Step 3

Hitch up a movable block behind the spoon to control the shooting angle. We looped the rubber band around the bottom of the block so it could slide back and forth.

Step 4

Use rubber bands to fasten it to the base board. You're going to want to slide it back and forth to control the shooting distance.

Step 5

Choose your projectile. We like pompom puff balls, since they travel slowly and don't roll too far, but marbles or paper balls or beans also work. Choose your distance, choose your angle, and shoot some hoops!

Check it out

- The ball always follows a certain path from the shooter spoon. What shape does that path look like?
- If you shoot a Ping-Pong ball, does it follow the same path as a puff ball? How about a small wad of paper?
- In terms of the path, what is the trick to getting a basket?
- How would you have to change this arrangement if you wanted to make it twice as large?

Make: Carnival Ball Game

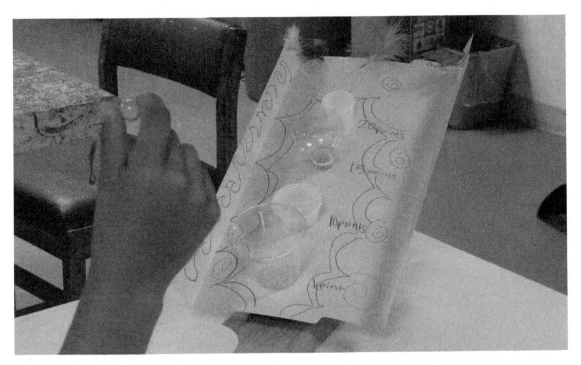

Figure 5-1. *Tinkering with the carnival ball game*

Gather More Stuff

- Another base board, similar to the first
- Another smaller piece, set at an angle
- File folder
- Small cups or bottle parts
- Rubber glove or balloon
- More rubber bands
- Marble or small ball
- Drywall screws (optional)

Gather More Tools

- Drill and screwdriver bit (optional)

Tinker

Again, it matters not how you slap this thing together. You could even set this whole thing up temporarily propped against a chair or something, with no fasteners at all.

Step 1

We happened to have a beautiful piece of oak trim, scrounged from a local woodworker, so we built ones on that. The other pieces are screwed in, but nails or even glue could be used. The angle shown here, around 30°, works pretty well.

Step 2

And again, the backboard is for ease of play; this one, made from an entire unfolded file folder, makes the marbles roll back to you most of the time. The launcher cup should be sturdy, and you can also mount it on an adjustable block as in the basketball hoop project. The other vessels can be anything you like in any arrangement.

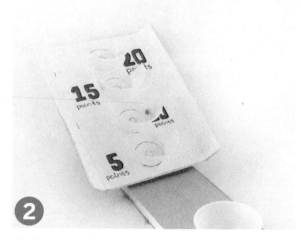

Step 3

The launcher here is really a bouncer. You stretch a latex glove or a balloon over the mouth of the cup, and secure it with a rubber band so it stays tight. Then you bounce your marble off this mini-trampoline and into the cups ahead. Biggest points for the smallest, farthest cups!

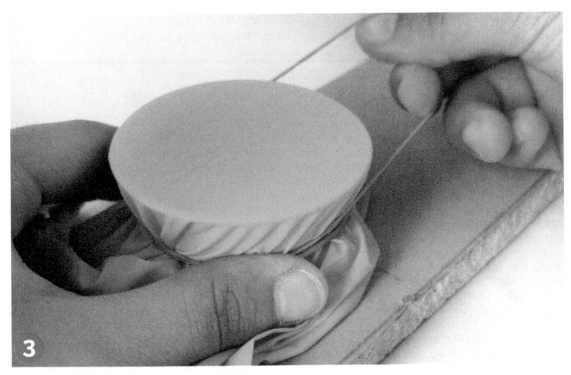

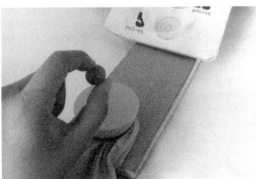

Check it out

- How is the path of this marble similar or different from the one in the basketball game project?
- With what factors can you adjust the final landing point of the marble?
- How would a similar sized steel bearing behave differently than the marble?

What's Going On?

Things that go sailing through the air without their own power—cannonballs, footballs, handballs, marbles—all follow a path called a *parabola*. A parabola looks a bit like a rainbow, except that a rainbow is half a circle, with the two ends coming back together at the bottom, which you rarely see in the sky but can easily see with a garden hose misting in the sun shine. A parabola never comes back together and is a bit pointier at the top. It's one of several shapes called *conic sections*. They are named that because you can construct them by carefully slicing up an ice cream cone.

So anything soaring without additional power (planes, rockets, birds, and bees don't count, since they've all got an onboard power source) will soon be falling back to the earth following this characteristic shape. There are a couple of exceptions here. One is if an object happens to be going straight up or down. A parabola must have at least a bit of sideways motion. Another exception is that if something hits a certain speed called *escape velocity*, and is heading upwards, it will never come down. It has freed itself from the earth's gravitational field. That velocity is around 25,000 miles/hour, so it's not likely to happen with your marble.

You can start the projectile from any point on the parabola. Think of standing at the lip of the Grand Canyon and kicking a rock straight out over the abyss (terrible idea, by the way: you could kill someone with that rock). The rock will follow a parabola, but only one side of it. In other words, it starts at the tippy-top of the parabola heading forward and then begins the graceful curve downward. If you should pick up a rock

and throw it down into the canyon, as opposed to straight forward, again it will follow a parabola, this time starting partway down one of the sides.

Skilled ball players are masters of the parabola. When your favorite sports hero makes a star move, sending the basketball through the hoop from half court, or delivering a Hail Mary pass into the arms of a running back at 80 yards in American football, or slugging the baseball over the left-field fence, what they have done in essence is to send that ball on a parabola that starts with them and ends exactly where they had intended. That is also just the way to win the two games you just built.

The interesting thing is that there is more than one parabola that will do it. In fact, from a given position on the basketball project, there are an infinite number of parabolas extending farther and farther up, each of which returns down into the hoop, swish. To get the different parabolas, you need to change the angle of release and the force given the ball. You change the angle of release by moving the little block behind the spoon, and you control the force of the shot by how far back you pull the spoon. Now, your spindly little spoon will give out before you can try them all (and it's always a bit of a chore to try an infinite number of things; takes forever), but you can try a few.

See if you can shoot your ball with a low, broad parabola into the hoop and also a high, pointy one, all from the same position.

If you're using a puff ball, you may notice that the path it follows is somewhat parabolic but not symmetric. It's flatter on the far side. This is because it has a huge amount of air friction and a very low mass, so the air slows it down quite a lot. The air will slow down a Ping-Pong ball and a paper wad as well, for similar reasons. Marbles will follow much more of a true parabola, since the mass-to-air-friction ratio is much bigger. Marbles and hard things are a bit more dangerous though, so be careful.

Sports players have to consider the air friction factor, and sometimes there is a wind factor, too, taking the ball in a certain direction. Have you noticed that when watching or playing a ball game?

In the carnival game, you've got two parabolas, joined at the bounce. The first one starts in your hand, and the second one ends in one of the target cups (if you're good). It's harder to analyze, since you have to use the fast-moving marble instead of the puff ball, but it's also more fun. You can envision a real crackerjack player winging that marble down into the glove, it rebounding up to the ceiling and back down—chup—right into the cup. All following a parabola.

The factors you'd have to master to get this to happen are quite a few. Some are the strength of your throw (which will determine the speed of impact into the glove), the exact point of contact on the glove (you get different bounces when it lands in different places), the angle of impact onto the glove (to a rough approximation the angle of impact will equal the angle of reflection, just like for light in a mirror), and the weight of the ball (a steel bearing will have around four times the mass as the marble, and thus receive a different acceleration from the glove, just as Uncle Isaac said). Tweaking all these factors I believe is what makes it so much fun.

Keep On

Pinball machines (*Figure 5-2*) and marble rolls are great mechanics projects as well. With these games you've still got gravity pulling, but the inclined plane with its obstacles designates the path. In the photo here, you can see the narrow launching chute on the right ending at the top in a curved path that lets the ball out left to fall

through the spinners, ramps, traps, and levers until it falls into a slot at the bottom. We've made a lot of great marble rolls with just nails or just chunks of wood glued on a slab, but the coolest ones we've made were when we could find some little bells for the marble to ding.

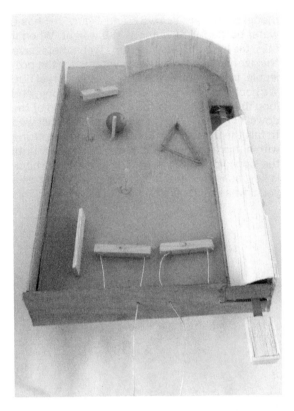

Figure 5-2. *A pinball machine*

Catapults and trebuchets are glorious subjects for tinkering. Tinkerer's heaven is filled with them, in all shapes and sizes. Catapults are powered by something stretchy; in *Figure 5-3* and *Figure 5-4* it's a rubber band. Two designs here launch the projectile when the throwing arm hits the stop. The throwing arm in *Figure 5-3* slides back and forth so that you can designate the angle you want to release from. (The throwing arm swings up between two upright dowels, stopping upon contact with the cross piece at the top.) The model in *Figure 5-4* can be adjusted, too, by tilting the whole base backwards. When it is sitting flat, it releases at 0°, straight ahead. With either of these two beauties you can reprove Galileo's mathematical calculation that a 45° release angle will lob your pebble the farthest distance.

Trebuchets use only gravity to hurl their projectile. This was a big advantage back in the heyday of siege warfare, since there were no good sources of enormous rubber bands back then. The model shown in *Figure 5-5* and *Figure 5-6* has a water bottle full of water as the weight. When it falls, the long dowel swings up. Trebuchets have trickier holding and release systems for the projectile. In our model, the projectile is a steel nut. We hook it on a string about 0.5 meters long, and loop the string around the hook at the end of the dowel (*Figure 5-7*). When the weight is allowed to fall, the arm accelerates up, pulling the string with nut in a graceful arc, faster and faster, until the loop slips off the hook. The angle of the hook is key, as is the length of the string. It may take you a while to get it all right. On a good day, this

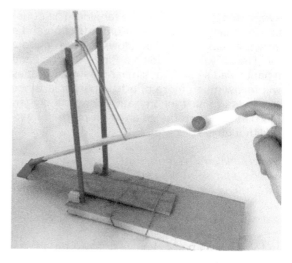

Figure 5-3. *The catapult*

ancient weapon can chuck the nut the full length of our basketball court (the real one, not the earlier project). If you've got tinkering blood in your veins, you're already scaling this one up in your mind, so let me tell you that the best time of year to make a giant one of these is right after Halloween, when you can get all the free pumpkins you want to send sailing across the field.

Figure 5-4. *A catapult with adjustable throwing arm*

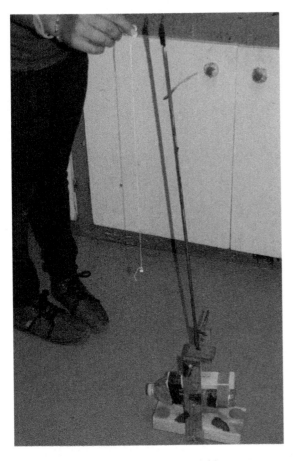

Figure 5-5 *A trebuchet with water weight*

Figure 5-6. *Close-up of the trebuchet*

And finally, if you're going to go to the trouble to launch something up, you may as well put a parachute on it so it will have a nicer ride back down (*Figure 5-8*). Parachutes made from cloth or plastic occupied a significant portion of my youth. I even made one out of my blanket in

second grade and was fortunate enough to let Mom in on my plan to try it out on myself from the peak of our roof. She managed to convince me that this particular project may compromise my entire future. I stuck with little action figures from then on.

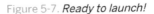

Figure 5-7. *Ready to launch!*

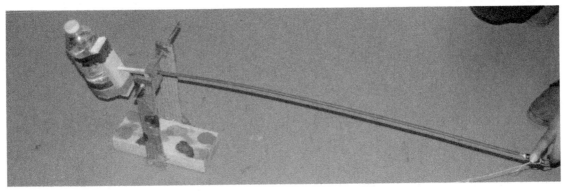

Internet Connections

- Base jumping looks like a real scream on YouTube, until you think about what happens when the parachute doesn't open right.
- Check out the siege weapons in the movie The Lord of the Rings: The Return of the King movie: search "Siege of Gondor, Minas Tirith." Oh, and it always help to have colossal ogres on hand to reload and cock your trebuchet.
- And have you considered putting this book down to watch a good hour or so of YouTube trebuchet launches? Look specifically for the floating arm trebuchet (FAT). These monster machines have an initial phase where the throwing arm together with the massive weight fall freely for 10 or 20 feet, building up speed and momentum before the rotation phase kicks in. Don't get in the way. It looks like some of them could bring a projectile close to escape velocity.
- In fact, you could blow the whole weekend searching things like "best basketball moment of all time" or "awesome football completions." Watch those parabolas: oooooo...

Figure 5-8. *Parachuting down*

Standards Topics Links

- Mechanics, dynamics, motion, movement, momentum, force, gravity, projectile, parabola, conic sections, simple motion equations, free fall, friction, and inclined plane.

More Tinkering with Mechanics

- **Mike Rigsby,** *Amazing Rubber Band Cars: Easy-to-Build Wind-Up Racers, Models, and Toys* (Chicago Review Press, 2007)
- **William Gurstelle,** *Backyard Ballistics: Build Potato Cannons, Paper Match Rockets, Cincinnati Fire Kites, Tennis Ball Mortars, and More Dynamite Devices* (Chicago Review Press, 2012)
- **Judith Conaway,** *Things That Go! How to Make Toy Boats, Cars, and Planes* (Troll, 1987)

6 Tinkering Logistics

You don't really have to make much of a plan to tinker; you can just dump the stuff out, let the students in, and start tinkering. But to optimize results and minimize crises and waste, it may be good to think about the overall logistics of how things will come together (or fall apart) as you tinker with your kids. In this chapter, I'll talk about the tinkering space and materials, the projects themselves, and then some loose ends.[1]

Tinkering Space

The essence of teaching with tinkering is to get tools and materials into the hungry hands of your students. Most educational institutions are not set up for this, so you may face a challenge to pull it off. The key is to see this as one of those magnificent, exhilarating challenges, such as learning to fly or summiting a tall mountain (as opposed to one of those miserable challenges like fixing your toilet when it's overflowing or salvaging a burned dinner for guests). I can assure you from vast personal experience that it can be a toe-tingling joy to present students an array of materials and basic tools in such a way that they feel freedom to use them safely toward a broad diversity of products.

To drive this point home, visualize a group of girls, say fourth grade girls who've more or less concluded that the essence of life is tied primarily to what clothes they're wearing and what their hair looks like. You have other plans for them. You outline to them the tinkering possibilities that will be available in your program, and one of them involves learning to use a drill press with a hole saw. They'll need this instrument to make a hole in their jewelry cabinet with music box, regardless of what form theirs will take, and you expect them to line up and learn to do it when the time comes.

You can see the girls catch their breath and look at one another with wild anticipation. This anticipation builds until they are moving steadily forward in the line at the drill press clutching their pieces of wood ready to be hole-sawed. Each one in turn reaches the front, dons the safety glasses, and receives your safety pointers, then grasps the handle of the powerful machine. They realize either from your pointers or from the feel of their own muscles that a large machine may not require a large force; gentle finesse on the handle is all that's necessary. Faces of focused concentration transition to sheer elation as the whirling bit bites into the securely clamped wood piece and slowly chews a perfectly round hole for their ballerina. As they walk with ear-to-ear smiles, you have no doubt that you have changed their lives and changed the world.

Let there be no ambiguity: it is 100 percent worth the trouble to make this experience happen for your students, especially the girls. At the very least, every kid deserves a chance to learn to use a hand power drill or some other power tool.

---

1. I won't be talking about how to set up your program as an organization: nothing on procuring a space, hustling funding, governance, insurance and liability, getting in good with the local political leaders, and maintaining a board that will support you and keep you in line. This stuff is critically important as well, but is covered in many books on the care and feeding of grassroots organizations. Some of our specific information can be found on *http://www.csw.org*.

Vignette: Noemi's Long Suffering Circuit

Something drew 12-year-old Noemi to put down one of her precious dollars and join the group of kids—mostly boys—making amplifiers from a printed circuit kit we'd designed and produced. After the third day of building, all the other kids had finished theirs, and she was still having trouble. She gave up, but I wouldn't let her leave. I told her it was against regulations, crazy, and impossible. Noemi is not one to back down easily, especially from giving up, and it became a bit of an ugly scene.

Something told me she'd come around though, so ultimately I got down on one knee, looked her in the eye, and told her she had started this and she could finish it and I was going to stick with her until she did. Veiled threat? Perhaps, but it worked. She rolled her eyes, dragged back into action, resoldered a few of the cold solder joints, and soon the gizmo crackled to life. The smile on her face as she thanked me on the way out the door will be with me forever.

Now, on to the details. It is best to have a real workshop to tinker in. Sports are a good case to compare to: no one would expect kids to learn and excel in sports without a proper facility. There is little debate or philosophical justification necessary when building a new sports facility because everyone knows the value of sports in the lives of kids, and everyone knows a proper space is necessary to convey that value to the kids. The world has not yet come to the conclusion that the same is true for tinkering, but I trust that we will soon come to our senses and conclude this, indeed. The fact is that the price tag for a reasonably well-equipped tinkering workshop is cheap compared to a sports facility of the same caliber. It is also quite possible to argue that the benefits are still greater.

The Community Science Workshops (CSWs) we have in California are ideal for any sort of tinkering. Check *www.cswnetwork.org* for more information and photos on how we set ourselves up for maximum-option tinkering. Essentially, we blend clever staff, tools, materials, work stations, project models, and inspirational hands-on exhibits all in the same room. When groups come to us, or individuals drop in during our open-door hours, everything is all set up and ready to go in dozens of different directions.

You can make one of these yourself in your school or home (the original one began in Dan Sudran's garage).[2] An old wood shop or auto body shop is often perfect for transforming into a tinker studio. But a big, empty drywalled room works as well— that's what we started with in Watsonville. Ideally there is a rough floor, not too nice, not carpeted, and the space is not too near other studious occupations, since tinkering can be quite noisy. Good ventilation is a must too, since there will be dust. In fact we haul our dustiest machine—a disk sander—out the back door for use. Any tables will do, but big, thick, strong ones are best. Plywood and two-by-four construction has served us well for years.

Ideally you let your clients in to help you set up your space. They may do more hindering than helping, but you'll immediately begin your relationship with them, and you'll be able to form

---

2. This is one way CSWs beat big-budget science museums: every community can have one, just as they have libraries and youth programs in art, music, and sports.

the program precisely around their needs and behavior. Sometimes timing doesn't work out for that though, and you have to cram a bunch of stuff together and run with it. That's fine, too, as long as you can remain flexible and keep a firm eye on the prize of serving the kids on their own terms.[3]

In recent years, hackerspaces and Young Makers workshops have sprung up like mushrooms after a rain. Many use common or semi-common spaces such as industrial garages, warehouses, or unused institutional space. Many are focused on technology, but others do some sort of craft or art or food creation. Some are primarily for youth, and others for all ages. I hope it's clear: there is nothing—not one thing —standing in the way of you and your buddies creating a community tinkering space for yourselves and your students.

If you're a teacher and can't find or create such a space, don't worry: you can do good tinkering in a classroom, too. It will take some rearranging, but it can start small and grow. You can't be afraid of *stuff*. Piles and piles of stuff. If you have an aversion to stuff, I suggest you go into English or history, where you can tinker with less stuff. To tinker with science, engineering, design, construction, electronics, crafts, or food, you'll need stuff and lots of it.

Let me show you around the Watsonville Environmental Science Workshop 16 years after its founding.

The first things you see

Just inside the door, among other things, you can see a sewing machine beside sewing books in front of sewing project models, a three-level rack, a shell and fossil display, a porpoise skeleton we found and cleaned off ourselves, the refrigerated salt-water aquarium holding local specimens under the counter, a microscope viewing table, and the cat pelt we tanned ourselves from a euthanized cat now tacked to the ceiling:

Storage racks and bins

Here. left to right, you can see some of our storage racks, for a variety of materials, our limited hand tool rack, and our small parts bins:

---

3.   Some of the worst science museums I've visited, incidentally, have begun with plenty of funding and everything in place on opening day. They missed the mark because they were focusing inward to their fancy exhibits instead of looking outward to the communities they were supposed to serve.

Here is a good section of our project models on display, around the central scroll saw and beside the hand power drill charging station behind the drill platform. Note the enormous exhaust fan for use when we're sweeping up:

Our knee-level library is visible to the left beneath the music section under electronics cabinets, all behind the central paint, glitter, and gak table alongside assorted exhibit tables:

Stocking Your Space

Project models and more tools

Surrounded by more project models, you can see the mandatory hot-glue station in front of more small parts bins, now mostly for bits of decorative detritus. The drill press stands just inside the back door, and welding shields hang above:

Having a modest supply of dozens of possibly useful materials on hand is integral to tinkering. I can't begin to count the number of discoveries and breakthroughs we've made in our Workshop due entirely to the fact that the perfect item just happened to be sitting around next to where we were tinkering. You're not going to realize that bobby pins work perfectly for the spring mount on the woodpecker project unless they're hanging around. You won't think to use a toaster heating element as a motor speed control unless that junk toaster is hanging around. Corks, clothespins, marbles, balloons, sugar, CD cases, bamboo skewers, string and wire of all sizes, marker and bottle caps, broken DVD players, pennies, vinegar, tiny lightbulbs, lenses, and key rings all take their places organically in your ongoing projects if they are readily accessible.

Scrounging and dumpster diving are good skills to have. I notice that even when I'm on vacation, I view a pile of refuse with an optimistic eye for possible tinkering materials. We also buy a lot of stuff from cheap household supply shops and from hardware or home improvement stores. We frequent the lumber yard, irrigation supply shop, and electrical supply shop. We've got a relationship with a local metalvworker as well as a number of cabinet shops. We do order some stuff online, mostly for convenience when we need a bunch (we buy hot glue by the 300 pounds). Each of the directors at the Community Science Workshops has cultivated relationships with industry and retail/wholesale suppliers. We've all got a line we use to get in good with them. Usually we'll bring a project some kid made to show them what we're going to use their donation for. No reason to get nervous about this: they'll either say yes or no, and it is all for the kids. You can tell within a few seconds whether or not you're wasting your time.

Manuel, director of the Fresno Science Workshop, and I often swap stories of chiseling for donations. He makes a point never to buy anything from any new source without introducing his programs and mentioning that he'll be able to serve a lot more kids on a meager budget if they can offer a donation or discount. He has even succeeded in getting two places to compete for who donates the most to him. To my knowledge, he's never been run out of a store by security guards, and he has brought in many thousands of dollars' worth of good, good tinkering treasures. One of his biggest coups was the local hospital, which tosses out hundreds of pretty good batteries each week, I guess since it is embarrassing to find a dead battery on a life-support system.

Getting steady donors makes life easy because they'll always have more stuff for us; waste is part of their system. The mediocre ones let us stop by and pick up their waste. We thank *them*, but in many cases we're actually saving them from paying to take it to the dump.[4] The best ones bring it to us when they have a load. In Watsonville, we even have a parent who takes requests and then hauls them to us in his pickup, with two preteen daughters helping him to load and unload it. It's good to have connections in institutions of learning, such as the local college or school district, which may be able to pass on outdated science kits or old lab apparatuses.

If you have a local museum in your area, by all means let them know you're ready to take whatever they're tossing.

There is so much waste in modern industrial society that it makes sense that you should think of your work as environmental in every respect. You're reusing stuff that would be taking up space in a landfill, and by doing so you're leaving other stuff on the shelf, which avoids the further extraction of raw materials from Mother Earth. You'll always have to buy certain items, but there will also be plenty of quality tinkering done entirely with found or donated junk. This reduces your dependence on a big budget, too.

In terms of financial support, the existing Community Science Workshops have all employed a strategy of diversifying funding and support across three arms, more or less as follows: public agency (city government or school), grant funding (public or private), and contracts (gigs for pay). With this model, one of the legs may collapse completely, and the program can still maintain some forward momentum.

4.   A bit of caution is advised on donations of hazardous material such as lead-acid batteries: you and your program may incur a charge if you have to go to a special place to dispose of it when you're done tinkering.

Materials and Tools

Here's a list of some of the stuff we have hanging around.

Materials:

- Kitchen and household stuff: disposable utensils, bottles, cans, cups and plates (all different types have their calling), bamboo skewers, toothpicks, clothespins, straws, paper, and aluminum foil.
- Office stuff: binder clips, paper clips (large and small), a stapler, rubber bands of all sizes, tacks, push pins, brass fasteners, etc.
- Small structural vessels: corks, film cans, salsa cups, and medicine bottles.
- Wheels: bottle lids, film can lids, checkers, poker chips, large beads, or wood circles cut with a hole saw.
- Structural wire: baling wire is useful, as are clothes hangers, picture hanging wire, wire twisty ties, pipe cleaners, and even bicycle spokes.
- Fasteners: nails of various sizes are used, and to make something stronger, you can use drywall screws driven with a hand drill.
- Wood: three standard sizes for light construction, paint paddles (for stirring a gallon of paint; these are free at the paint shop, so grab a handful!), tongue depressors, and craft, or "Popsicle," sticks. For beefier projects, 1x2 furring strips are quite useful, and a small pile of scrap wood from 1/4" to 3/4" thick is always in demand for base boards and other parts.
- Other materials: plastic of various types (Plexiglas, bubble wrap, Styrofoam, CD cases, plastic bags, and random scraps), sheet metal, cardboard thick and thin, cardboard tubes, funky metal hardware, pipes of steel and copper, PVC, rope and twine, old restaurant menus, and anything we find that's not particularly dangerous.
- Motors: buy one or two from RadioShack or order a dozen from Kelvin Educational (*http://www.kelvin.com/*). Rip them from old toys.
- Batteries: cheap batteries at 99-cent stores don't last long, but that's mostly what we use. If you screw up and leave a project connected, it will drain a high quality battery just as completely as a cheap one. The best is to invest in a charger and use nickel-metal hydride (NiMH) rechargeable batteries. They'll last for years if you take care of them and so are cheaper than any other type in the long run.
- Electrical wire: thin copper wire from phone or Internet connection is what is used on most of our projects. You could also use bare copper wire, but if it contacts another wire, you'll have a short circuit. For electromagnets, you'll want magnet wire of 24 to 30 gauge. If you just get one size, get 26 gauge. You can get it at Kelvin Educational, a motor repair shop, or by ripping apart an old transformer or motor. For the speed control, you can find an old toaster and rip the heating element wire out of it; connect one end solid, and run an alligator clip along its length to control the current to the motor.
- Decorations: paint, markers, glitter, ribbons, colored paper, stickers, puff balls, pipe cleaners, etc.

Tools:

- Drill with bits
- Files
- Hammers
- Hot glue guns with plenty of glue sticks (low temperature works fine and is safer, but high temperature bonds a bit better)
- Knives: box cutter, hobby knife, razor blade
- Needle-nose pliers
- Pliers
- Rulers
- Saws, wood and metal
- Scissors
- Screwdrivers: flat and Philips, large and small
- Side cutters
- Tape: masking, black (electrical), duct, or gaffer's
- Vise or C-clamps or the long ratchet-spreader type sometimes used for wood gluing

This basic element of procuring a wide array of materials is important on many levels. Tinkering is inevitably linked to understanding the properties of various materials. Kids need to fiddle around building things from various materials in order to understand why certain things are always made of metal (saws, nails) and others always of plastic (straws, wire insulation), while still others can be made of either (pens, bottle caps, and cups). Thus you need to put those materials in front of the kids.

In addition to materials, you'll need to procure tools. Each project in this book has a list of tools, and you'll likely find yourself wanting others.

Avoid buying them all new. That's more expensive and also cheating—did you remember this is an adventure? Start with garage/tag/yard sales, thrift shops, and pawn shops. Ask all your friends and relatives for tools they undoubtedly have rusting away in the garage. Go to hardware stores and look for the discount or reject bin, then ask for donations and special deals (for the kids!). Go to cheap household supply stores. Don't buy something that's so cheap it's dysfunctional—for example, an astonishing number of side cutters out there don't actually cut—but don't spend any more than you have to in order to get basic stuff.

Vignette: Tales of Heroic Substitution

We've got many gripping stories of laying out materials in front of, say, 30 screaming sixth graders and realizing we've forgotten one of the key items. Sometimes we're able to call for reinforcements, but other times we can't. With desperation as motivation, we dig deep into creativity and sometimes end up with a significant breakthrough that may be better than the original design. Here are a few examples: If you forget your connection wires, you can use the steel wire that winds together a pipe cleaner. If you forget your base boards, you can use three layers of cardboard glued together. If your drill quits, a lot of times you can punch or hack the hole with a hammer and nail. If you run short of nails for the xylophone project, you can just put four at the corners, suspending the entire set of keys like a musical hanging bridge. Glitter can be made by scissoring up chip bags or whatever colorful paper trash you've got hanging around, and a chunk of Styrofoam is also reducible to a glitter-like substance. Soap and lotion will sometimes have enough color to substitute for food coloring, and also make the liquid thicker. Pencils substitute for dowels, and don't forget that the very trees outside your edifice may have small sticks or leaves perfect to enhance the functionality and/or aesthetics of your project.

Immediately after procuring tools, you have to worry about kids stealing or breaking them. Quality tools tend to disappear more rapidly than cheap ones, and we've generally seen that students break quality tools just as fast as cheap ones. There will always be attrition; just try to keep it to a minimum. (And anyway, it's always nice to imagine the kid who lifted the side cutters using them to build and fix more great projects at home.)

As you amass these tools, begin using them yourself at once. You have to be sure both that they work and that you know, more or less, how to use them. They must be safe, that is, they must function in the intended manner without hazardous breaking or slipping.

Begin thinking now about your storage spaces. You'll want a bunch of boxes for supplies both random and regular, and at least one tool box for the tools. Label things as well as you can, and try to keep it organized. The minute you can't find something, it's just as if you don't have it at all. Certain items will be used frequently, and it can be worth buying some of those little plastic sets of drawers to have it all accessible. Go to *www. instructables.com* and search "workspace" to find many clever ideas on how to organize and stash a lot of stuff in a small space. For us teach-

er types, it is one step more complicated, since we want our students to have full access, but at the same time put reasonable limits on how much stuff they can snag. In our mobile Workshop van, one set of plastic drawers is labeled, "Adults only." Therein lies minor contraband (balloons, rubber bands, and razors) and our high-end, highly sought after items (magnets, motors, fancy tools, etc.). We still lose stuff from there from time to time, but it's a reasonable compromise.

Figure 6-1 shows supplies for the mobile program all jammed in the back of the van.

Figure 6-1. *Our stuff, packed*

Vignette: Emilyn's Tinkering System for the Satellite Sites

We've taken projects to satellite sites since the first year of operation at Watsonville Environmental Science Workshop, but in general it was parts and tools for a single project or two. When we got funding from a heaven-sent anonymous donor, we immediately began preparation to serve outlying sites in high need of student activities. Emilyn was a recent addition to our permanent staff and was convinced, together with myself, that we could offer these sites the full-option, full-energy, full-chaos tinkering that we do at our main site out of the back of an old city van. She designed simple cabinets and storage systems that we slide in and out of the van with extreme efficiency, and then began building relationships with dozens of the denizens of these low-income sites. Within months, it was clear that not only is it possible, in the long term, this scheme is more sustainable than presenting single projects. Kids of all ages can find good stuff to do, stuff they've never done before, and continue learning each week in their own way, at their own level.

Figure 6-2 shows our good assistant Nestor lugging the stuff in and out of the spaces we use. We tried various wheeled mechanisms, but in the end it turned out to be easiest just to grab and lift.

Figure 6-3 shows one of our three sets of plastic drawers (all of which we salvaged from the dump). This one holds the contraband—rubber bands, batteries, motors, and magnets; anything the kids are likely to want vast quantities of—thus the red cover cloth. We place it off to the side for easy teacher access but more difficult raiding by kids.

Figure 6-2. *Lugging our stuff*

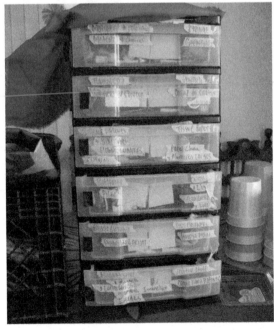

Figure 6-3. *A drawer full of consumables*

Figure 6-4 shows Emilyn's famous drawer-keepers, installed with hot glue on the plastic sets of drawers. These have saved us hundreds of hours of picking up spilled supplies in the process of hauling, unloading, and loading.

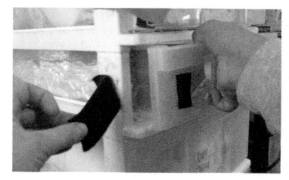

Figure 6-4. *Emilyn's drawer-keepers*

You can see the entire suite of supplies for our mobile program in *Figure 6-5*. Still in the van are a drill press, more scrap wood, a few other boxes with project supplies (gak, papier-mâché, electronics/circuits) that we'll haul out if necessary, as well as six empty seats to carry students around if necessary.

Figure 6-5. *Mobile program supplies in the upgraded van, which we got after the city saw that the program was working great*

Tinkering Projects

A project is something you focus your tinkering on for a while, and often something you can take with you when it's done. You can do a lot of tinkering without ever having a finished product, but often you will have one. I've laid out some characteristics about our projects earlier, and I include directions to some of our best projects in this book. Here are a few additional pointers about the projects our kids tinker with.

A collection of books outlining projects that are doable with the supplies you have on hand is a good thing to have. Also, a collection of photos of previously created projects is inspirational. But the most valuable of all is a great number of finished projects, in working order (more or less), hanging from the ceiling or on small shelves on the wall. Kids can take them down, find out what they are all about, and then use them as models for their own version.

(Kids often break the models again and again, and sometimes steal them, always the best ones. Some of this is inevitable and not such a big deal, but it takes a lot of staff time to fix and remake them. Some loss can be avoided by putting the projects on higher shelves and hung from the ceiling.)

Once a functioning model is on the table, the work of the staff is reduced greatly. No one needs to give step-by-step directions. Kids can copy the model directly or develop the project in different directions. We often encourage our better tinkerers to improve on the model, as opposed to just copying it. Teach us something!

Here are some popular projects that we always have on display in our Workshop and that we are continually fixing and refixing. If you're not familiar with one, look for it on our website (http://cswnetwork.org/activities/):

- Tornado in a bottle
- Electric cars
- Birdhouse
- Magnet hanging up
- Two-by-four scooter
- Saxophone/membranophone
- Hydraulic car or butterfly

- Catapults, onagers, trebuchets
- Toilet demo model
- Jewelry box
- Air and water rockets
- Gak—the amazing polymer toy, also known as slime
- Bean bags and handbags (near the sewing machine)

Vignette: The Moose Call's Developmental History

The membranophone has always been a popular project. It's sometimes called a saxophone, since it can sound remarkably similar, but instead of a reed, it's got a vibrating balloon driving the sound. You blow into a small bottle (we used a film can in the old days), and the air is forced to stretch the balloon and escape up over the edge of a PVC tube. Years ago, someone at the San Francisco Mission Science Workshop realized you don't need a container to make the effect, but can instead tape the mouth of a balloon onto the PVC tube, and then tape a straw into a small hole in the top of the balloon. You then stretch the balloon sideways over the end of a tube and blow. (Kids will miss the critical stretching arrangement at first, puffing away with no sound coming out, and we rarely tell them the answer, but rather ask them to observe carefully as someone makes one work.)

Then came the inevitable day when we were set up to do that project with a large, boisterous class of kids and realized that we were out of balloons. Modesto Tamez summoned all his experience of 20 years teaching science through tinkering and reached for the latex gloves. As the class chopped them up, he realized that the cost and sound of balloons and gloves are nearly equivalent, but there are five fingers on each glove, each able to fill the vibrating role of the balloon, thus reducing the (already negligible) cost of the project by a factor of five. If there was a Nobel Prize for tight-wad science teaching, Modesto would have won it many times over. Lessons so far: desperation is the mother of invention; and never be satisfied with how cheap a project is—you can make it even cheaper.

Recently, we did this project—christened the Moose Call for its pleasing melodic output—at an Earth Day festival. We had several high-school assistants helping the public to construct these and saw around 300 of them walk off gaily blaring. It was a barebones construction (*Figure 6-6*): tube of donated scrap sprinkler line, half a straw, one latex finger, and black tape.

Later Gustavo and I worked out how to hot glue on a 45° wooden triangle (*Figure 6-7*) to one side and then glue the straw to it, thus enabling kids to play it with only one hand. At the next festival, a smaller one, we served the kids up with the additional wood block and a paper cone at the end, adding volume and directionality. This just about doubled the time it took to make them but added richness and gave rise to more questions.

The next week, I took a Moose Call model to one of our satellite Workshop sites, and 10 more kids made them. Now there was not the artificial, urgent time restraint of a class or festival, so they made them larger, more grand and ornate. And louder. We drilled two or three holes in the side to enable a change in the moose's pitch, and some even realized that two glove fingers together on two tubes (*Figure 6-8*) would produce a beat frequency[5] and deliver the familiar harmonic tone of an 18-wheeler bearing down on you. More lessons: different environments may call for different approaches with the same project; and even when you think you've developed a pretty good project, you may not have even scratched the surface.

Figure 6-6. *The moose call*

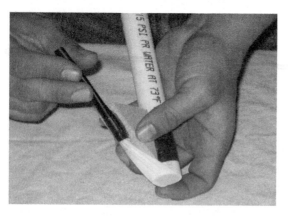

Figure 6-7. *The wooden triangle secures the straw*

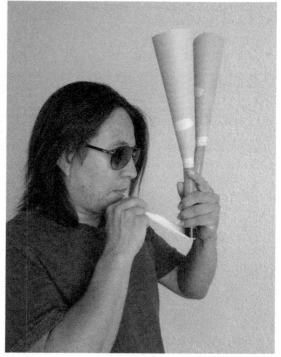

Figure 6-8. *The double-tube Moose Call*

We are often asked where we got the ideas for all these projects. All of the staff at the Watsonville Environmental Science Workshop have come up with great projects from various sources of inspiration, but my assistant Gustavo and I are somewhat obsessed. We've got so many ideas waiting to be tried that we're often found sneaking to the Workshop early Sunday morning or late Saturday night, just to get time to try one out. New ideas are the least of our problems. He and I have been hatching ideas like this since before we were potty trained. Asking us where we get ideas is like asking a peacock where it got such nice colors on the tail feathers: beats me!

However, Gustavo and I both understand that just like in music and fashion, nothing, but *nothing*, is really an original idea. We're all just stealing and trading around the same themes and trends

5.   When two sounds are close in frequency, they superpose to create a third, much lower frequency called the beat frequency. The beat frequency sounds more like a texture or warble in the other two sounds than a separate tone. You can make it by whistling with a friend: have them whistle a single note, and then try to whistle the same one. As you approach your friend's frequency, you'll hear the beats.

and sometimes—BLAM!—something wonderful pops up. We get a lot of ideas from watching students and seeing their successes and failures. We surf the Web. I flick through project books at libraries and book stores like a goat looking for blades of grass amongst the rocks. 'Tavo watches how-to and MacGyverlike shows on TV, gleaning what may actually be workable.

We focus less on finding new project ideas than on making the ideas we have come to fruition. No sooner do we catch wind of an idea than we're methodically thinking through the materials and parts at our disposal, deciding what to try first, what possible scale and structure to give the project, and how to assemble it in a doable way. We've both lost sleep thanks to pregnant project ideas. We were smiling wide though, as we lay there staring at the dark, minds churning with magical images of fantastic projects.

What is it about our best projects that makes them so good? Somewhat hopeless to nail down, the question is still worth thinking about. Here is a short list of factors I think play a role:

- Project is exciting to interact with. We find projects that throw or shoot something (stopping just short of being a weapon) and projects that make lots of noise are two of the most popular categories.
- Project works even though you do a shoddy job putting it together.
- Project is not too easy, and not too hard. Challenging but doable.
- Project is doable in a single visit. In the earlier project list, only the scooter is a multiday project. We continually encourage kids to do more complex projects, but in the end, the most popular ones are immediately gratifying.[6]
- Project is done with mostly recycled stuff. Kids are in awe that something cool can be made from junk.
- Project core is easy to duplicate, but finishing touches can lead to much variety between projects.
- Project is appealing to both younger and older kids, as well as adults.
- Project demonstrates some phenomenon that can be clearly and compactly articulated.

Vignette: Paula's Pump Evolution

I was driven for years to build a small pump to make decorative fountains. I made a ceramic fountain back in high school and purchased a little $3 pump for it. Now they cost nearly $20. I went through multiple prototypes, slowly simplifying it to the point that most kids can now build one, made from a 60-cent motor with a hot glue stick as the impellor inside a film can (described in my book, *Kinetic Contraptions*). But along the way, Paula Smelke at the Mission Science Workshop in San Francisco was fiddling around with my prototypes and strayed from my sacred goal when she noticed that the kids were impressed by the water just going around and around inside a bottle. Some called it a blender and others formed it into a toilet, known to be a very popular subject for kids of all ages. The final arrangement that evolved from that bit of serendipitous observation has become one of our most popular projects ever: a motor driven tornado in a bottle (*Figure 6-9*).

6. This question of immediate gratification gets at a much deeper issue that fascinates me. I often sermonize to reckless students about how they are not likely to die soon, and so it would be best to have a plan, at least for the next few years, a vision or dream for the future. At the same time, I've studied the slightest bit of Buddhism, and one of the main points is that *right here, right now* is really all there is, and vastly more important than anything past or future! So are our rough, delinquent, life-on-the-edge, hell-for-leather, live-for-today students actually latent Buddhists underneath it all? I'm not sure, but I'm led once again to believe that I have many, many things to learn from them.

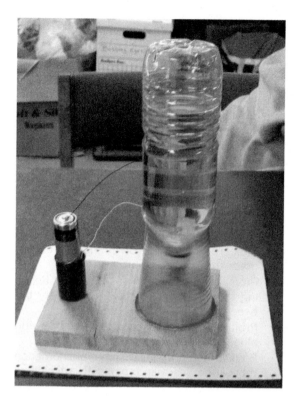

Figure 6-9. *Tornado in a bottle*

tiny but very secure pivoting system.

- A paper clip makes a good electrical connection when clipped onto aluminum foil in the way it was designed to be clipped onto paper.
- A cheap hobby motor runs with one 1.5 V battery and even better with two.
- The motor, battery, and some sort of switch can all be mounted on a tongue depressor and used interchangeably. We call this the Roach.[7]

There are dozens more, but you get the point: being familiar with a broad set of low-tech elements made from supplies you have on hand will get you farther when you have a new idea.

When building with junk and cheap stuff, we often find that certain parts fit together well. We call them low-tech elements, and make wide use of them from project to project. Here are some examples:

- 1/2" PVC tubing fits snugly into the mouth of a standard plastic drink bottle.
- A C battery fits nicely into a film canister. Not that you can find many film canisters anymore, but hey, it was good while it lasted. The next generation of this one will have something to do with a USB cord, I think.
- A segment of hot-glue stick with a nail hole fits tightly onto the shaft of a motor, thus making a connection with impellors, propellers, and rubber-band belts.
- The same hot-glue stick segment also skewers onto the point of a pushpin, enabling a

---

7.   I wrote a book about this arrangement and others: *Kinetic Contraptions: Build a Hovercraft, Airboat, and More with a Hobby Motor*, Chicago Review Press, 2010.

Vignette: Newton's Cradle

We worked for years on the ideal way to create Newton's cradle (Figure 6-10), the line of balls hanging from two strings each that clack back and forth when you raise one to the side and release it. You can buy them from a specialty catalog for tens of dollars, but we wanted to make one for cheap or free! The key was continuous adjustability; since no one could build it perfectly, it had to be possible to keep tweaking it forever. One by one, the low-tech elements formed into a highly doable project, even for younger kids:

- A large base board to keep the whole thing stable.
- Parallel vertical paint paddles onto which can be clipped tiny binder clips. This gives the balls three dimensions of variability.
- Large "shooter" marbles hot glued onto thin string have more inertia and thus more forgiveness of error than small marbles. They are cheaper and more accessible than metal balls of that size.
- When the hot glue comes unglued, the hot glue depression still stuck on the string can be reglued to the marble with a drop of superglue, making a much stronger bond.

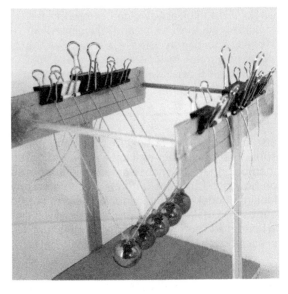

Figure 6-10. *Newton's cradle*

Facilitating Projects

As I mention elsewhere, we take as our (at times overwhelming) challenge to make each kid's project dream come true. The first few minutes can be the most important. We often find ourselves twisting or tempering, honing down without snuffing out, offering alternatives we've seen to work without coercion to follow them. In the end, the kid will be most satisfied if the outcome is successful, but we want the kid's heart and soul and creative essence all over it, too.

Over the years, I've increased my ability to predict the chances of success on a new project idea. We do our darndest, but sometimes the universe, with its laws of physics, does not support a given idea at all. As a facilitator, you have to make a call sometimes as to whether you think a given project is ever going to work, and you have to communicate with kids about this issue from the beginning.

I'll often be frank but gentle, saying something like, "*I think you're going to have to change something fundamental in this idea if you want it to work with the tools and materials we have here.*" Most kids can handle that, especially if it comes together with suggestions on what could be changed. Other times I'm not entirely sure it won't work, but I'm out of ideas and I've got other kids to help. Then I'll say, "*I'm doubting that this thing will fly with your current plan. Don't let me stop you if you want to keep trying, but I'm going to go help those kids over there.*"

Certainly if there is some part of the project that you're unsure about, have the kid do that part first. This may seem counterintuitive, but makes sense upon inspection: "*It won't help to have the rest of the project done if you can't get that hard part to work, so try it first!*" Finally, I sometimes just give out a friendly warning: "*Brace yourself for that thing not working: I think there is a good chance it won't, but let's give it our best shot.*"

Vignette: The Impossible Scooter

Just because you can't see how it is going to work doesn't mean you shouldn't give the kid encouragement and all the ideas you can muster. Some of our best projects are the direct results of kids pursuing an "impossible" idea. Twin seven-year-old girls pestered me for weeks to build a scooter. The popularity of scooters had just spiked again, but these girls didn't have the money. I assured them that we can't build a scooter here: too complicated, we can't weld aluminum, they're too young to weld at all yet, we don't have the wheels, and think of the complex steering mechanism! They wouldn't take no for an answer though and suggested that we make one from wood. Exasperated, I eventually grabbed pieces of two-by-four and tried to explain why it just wasn't going to work. Then we all began to think about it, came up with a few ideas together, and saw some hope. Over the next week or so, these twins and I developed our two-by-four scooter. We found surplus wheels and used scrap two-by-fours, eye screws for the steering, and scrap PVC for the handlebars. We have now made well over 400 of these, and they work so well that kids come back to beg for more wheels from us when they've worn them completely out. One girl even decided she had to have a side car for her kid brother.

In other circumstances, you may have a kid who can't think of exactly what she wants to do, or has a few ideas but can't envision the whole thing. In this case, the best plan is to start something: get a few of the materials she thinks she may use and start mucking around with them. Incidentally, I'm wary of any engineering, tinkering, or "invention" program that starts out with kids in their seats drawing plans with pencil and paper. As mentioned above, many kids these days lack experience mucking around with much of anything, so the first step is to offer an opportunity for that. Often the vision of the project comes quickly once hands are exploring materials.

Let me interject here the importance of art in the world of tinkering. Of course you can tinker with just art alone, but generally, science is the background in which tinkering is visualized, and engineering or technical skills of some sort are often what students walk away with. But art is also a brilliant part of tinkering for various reasons. First, it is a fact well known to all our staff members that kids choose project models based on how good they look. If you put up a crappy looking model, nobody wants to build it. We do our best to make sure the project model is full of art and make sure there are decoration supplies in each box that goes out to our satellite sites.

Furthermore, kids may never have the opportunity to tinker with tools and construction in their formal school experience, but there is almost always some thread of visual arts in school. Often it is amateurish, often the activities are random and only surface level, but art is present nonetheless, and the important thing is that most kids have had a happy, fun experience with it. Thus, if you embed art in your tinkering, you'll make the kids feel at home; they'll see a fun thing they've seen before.

Finally, when the project is "finished," that is, it's doing what it was created to do, the art of decoration can give an opportunity for the student to continue interacting with it. In this way, they'll have more chances to observe it and understand it, as well as to make it their own unique masterpiece.

It's worth noting that art and science are not at all opposite ends of some sort of intellectual spectrum. Both artists and scientists need to have keen skills of observation. Both want to understand and describe the world. Both seek extensive information and diverse perspectives on the subject they are studying. In the end, many of the skills of an artist are quite valuable to the scientist as well. In short, kids needn't choose; they can be great at both art and science.

Whether in the free-form or class group arrangement often you'll end up giving directions on how to do a project. It's best to give *vague* directions. Every day kids are taught to read or listen to and follow directions, so why make them do it again when tinkering? Another great mentor of mine, Modesto Tamez at the Exploratorium Teacher Institute, has developed this to an art. He points out that giving less-than-specific directions has the wonderful result of a variety of product, which in turn offers superior learning opportunities. Which do you want: a class full of kids all with identical projects, or a class full of kids each with their own fabulous creation, all learning from the uniqueness of the others? If you want the former, tell them all—step by tiny baby step—exactly how the project is to be done, and stand over them with a stick to make sure they do it. If you'd rather the latter, give vague directions.

Often there are two or three key points about a project that kids should be shown in order to have a decent chance at success. The rest they'll be able to figure out from the model, or make up on their own. *This* opportunity is what they're *not* getting in school, so be sure and give it to them through tinkering.

A related point to never forget is that probably no more than half of the kids are ever listening to what you're saying. Sorry, teacher types, but I think maybe you've long suspected as much. If you're counting on the lecture format to transfer information, you're sunk. If, on the other hand, you're counting on the kids learning from the process of tinkering with the stuff in front of them, with you facilitating from behind, it may not make a lick of difference whether or not anyone is listening.

After directions are given and kids are tinkering away, the two most common comments we hear from the kids are these:

- "What do I do now?"
- "It doesn't work."

You can address the first by following the advice I give further on in the chapter on questions and answers. A fine answer is, "Yeah, what do you do now?" as you take the kid back to the model to check out what is done already and what is left to do. This process is absolutely critical. By asking what to do now, the kid is making a plea, saying in essence, "All my life I've been given step-by-step directions, and now I'm unable to move forward without them." Resist giving them more directions; now is the time for them to start learning from the model and weaning themselves from directions. There is nothing wrong with being a foot soldier or a factory worker, but kids should also be able to figure out things for themselves.

The second comment has a snappy answer I never hesitate to give if the kid looks like he's stable enough to take it: "Of course it doesn't work! *Not working* is the natural state of affairs in the universe. *Working*, on the other hand, is something precious and extraordinary, something that takes skill and persistence to achieve. Now how can we get this thing to *work*?" I also point out that troubleshooting is an everyday job for many, many occupations, from computer programmers to plumbers. We did our best, but it's still not working. Yet. Now the new task is to figure out what went wrong. Finally, I don't usually point it out, but never forget that this kid is set up for an enhanced opportunity to learn, since he'll end up looking more closely at how the gizmo functions. Read why mistakes are critically important in the next section.

When a kid is finished, you can always commend him for a project well done, but it is also nice to offer the kernel of a suggestion that perhaps he could do *even better*, be more creative, get better results, and achieve better quality if he were to keep working on it, or do it again from the start. This is, once again, something familiar to all artists and artisans: practice makes perfect and hones skills. You have to be the judge when deciding to put out that kernel or keep it for next time; take into account the kid's frustration and patience levels, the time available, and whether or not he'll be returning or has resources to keep going at home.

A Few More Tinkering Considerations

By the way, lists are basic to any tinkering operation. When we come up with a new tinkering activity at the Workshop, the first thing we'll do after taking a few photos is to make a materials list and a list of the various things we noticed. I feel it's more than appropriate then to make this final smattering of suggestions a bullet point list.

- **Safety is first**. Sounds like a cliché, but it's really a statement of reality. Little learning will happen and little fun will be had if someone gets hurt, so take your time and make sure your space and your program are safe. Here are some key safety precautions we take at our program:

  - No kids under 6 are allowed to participate without an accompanying adult. (This is due to the potential danger of random materials and tools available to all at the Workshop. In a more controlled space, you can tinker with kids much younger.)

  - All electric tools are locked except the scroll saw and the hand drills. We've seen it nigh impossible to sustain more than minor scratches from these two devices. (Don't tell anyone, but you can actually put your finger on the scroll saw blade without getting cut; it's just going up and down. Rotary saws are a much, much different story.)

  - Safety glasses are required for all electrical tools except the hand drills. A few

hand tools should also be used with safety glasses: cutters that send tiny bits flying and hammers on metal, stone, or other brittle materials.

- Only one kid at a time can use the electrical tools. We've had two minor injuries resulting from two kids "helping" each other on the scroll saw.

- No horseplay, running, biking, rollerblading, scootering, or skateboarding in the Workshop. (You'd think that'd be obvious, but then you're probably over 12 years old.)

- Small injuries will happen as your kids tinker, but with good planning and structure, big ones never will. We've had Community Science Workshops running in five Californian cities for over 20 years now and never had worse than a few deep cuts and stitches. We never hesitate to compare our Workshop to sports programs, especially for the liability watchdogs. Everyone knows that if you play sports hard, once in a while you get hurt. As long as all due precautions are taken, it's part of the game, and it's worth it. The same goes for using tools and fiddling with all sorts of materials. Only we've never had a broken leg, never a torn tendon, and never a concussion. (Heck, at my old high school, they used to park an ambulance at the end of the football field during a game, and my buddy still limps from an injury sustained during one of those "school spirit" events. Yeesh.)

On the bright side, don't forget that one can learn a lot from an injury: how the skin works and heals, how burn pain is different than cut pain, how blood pumps, how fingerprints can come back, how a blister functions, etc. Also, a good group debriefing after an injury can help raise awareness and avoid more.

- **Engage parents and families.** Some of the best golden memories we have at the Workshop have been families tinkering together. Sometimes they'll work on the same project, sometimes several different ones. It's lovely when the members of a family all feel comfortable tinkering and can pass an afternoon together that way. Even if only one adult and one kid can get it going together, I get the feeling the world is a better place. I've seen research that specific parenting techniques are less important than that one-to-one relationship between parent and child, and there's nothing like tinkering toward a shared goal to cement a relationship. Par-

Some parents are a pain in the butt, though, and you'll need strategies to keep them out of the mix. But here again, other supportive parents may be just what you need in this situation.

Vignette: Guillermo's Mom

Guillermo's mom followed him in one day to build a shelf, and when that came out well, she came back and built a bed, and then a chair, and then a china cabinet. Guillermo wasn't that interested in helping, but he was a mama's boy, so whenever she got stuck, he'd go put her back on track. He was just happy that she took him to the Workshop so he could work on his own projects: cars, boats, and airplanes.

Vignette: Sarah's Wild Brother

Sarah's kid brother Daniel was a real handful: loud, hyper, unfocused, and unmoved by logic. To make it all the more exciting, he had a speech impediment so that you could never be sure what he was trying to communicate. Fortunately, his sister was an angel. Sarah would bring Daniel in, bounce around with him for a while until he settled into something, help him for a bit, and then sneak quietly off to her own project, always ready to dash back over and rescue him from whatever pit of trouble he was about to fall into. She always spoke very highly of him, expounding on his many talents and attributes. When the Workshop staff and I saw this unique situation, we'd put our high-school helpers on helping Daniel the best they could so that Sarah could get her project done. Daily we'd see them walking off toward home together, comparing notes on the tinkering projects they'd just done. She was in fourth grade and he in second.

ents can also bring resources to a tinkering space: tools, materials, ideas, snacks, and extra sets of watchful eyes and helping hands.

- **Process is more important than product**. Sometimes kids will finish tinkering on a project, realize they are dissatisfied with it, chuck it in the trash and walk away. Don't worry: they still gained from the experience. While we'd all love to see completed tinkering projects proudly hung on the wall, it's just not going to happen all the time. Most finished products will be mediocre (by definition), and I can say from vast personal experience that only a few will be worth keeping long term. But my experience with all the duds was just as valuable as that with the keepers. Which leads us to the undeniable fact that...

- **Mistakes are critically important**. You want your kids to achieve success, but you also want them to reap the benefits of failure and repeated trials before that success. The kid who fails a time or two before getting it right will walk away with more knowledge, more skills, and more confidence than the kid who nails it the first time. Everyone in the room must be aware that mistakes are part and parcel of the tinkering process. Indeed, a mistake is an opportunity for the student to advance developmentally. You as the facilitator hold the key to making sure they make that advance, and don't just walk away all bummed out. It's primarily a matter of communication and encouraging a broad vision.

- **Know when to step aside**. Now hear this: everything I lay out elsewhere about supporting kids tinkering, dealing with questions, dishing out answers, working with difficult kids, scaffolding, etc. can be put on hold if everyone is tinkering along fine. Remember, one of your goals was to get the kids to tinker independently of you, and if they've got it, you can move away. Not too far: you can start up a side project related to the one they're doing, do a bit of organizing, fix a model; anything you want so long as you stay active near their tinkering. (It is a

bad idea to sit down and read a book—you're the leader, and they'll follow.) You can keep up a conversation with them to monitor how they're doing, but don't think you have to be in their face all the time. You are important, but your role is facilitator; they'll learn mostly from the stuff.

- **Don't be preoccupied by this, but take a stab at connecting students' observations to theory in the canon if you can.** In high-school and college lab courses, you often progress down a narrow experimental path toward a specific result that illustrates a theory you've learned in the textbook. This is good and indispensable in the process of coming to accept the theories as correct. When you tinker, you often don't have as well controlled an environment, but you can often make and agree on observations that lead to conclusions that align well with a given theory.

An example is the magnet activity in this book: play with magnets then read magnetic field theory; your observations will tidily match the theory. Another is the common "gak" activity with white glue and borax mentioned in the chemistry activities: you can't really prove that those molecules are polymers, but with a big enough lab you could, so why not tell the kids and get them to start thinking about it?

Many education groups have the explicit goal of turning kids on to science and engineering. This is only a peripheral goal for the Community Science Workshops, our primary goal being to facilitate kids discovering their world through direct contact with it. But still, while they are blissfully tinkering away, we don't miss a chance to say, hey, this outrageously fun thing you're doing now is essentially science and engineering. Basically, we're cheating them if we don't.

- **Whenever possible, make connections to careers**. I wanted to be a physicist since I was in sixth grade. Problem was, I was unclear about what a physicist did on a day-to-day basis. I was enamored by various biographies I'd read, some unrepresentative (to the point of downright deceptive), and I was in college before I got the full picture. Now I do all I can to let kids know that what they're doing while they tinker is what some people do professionally all day long and they get paid for it, too. We often say something like, "As you design and build your gadget, you're *being* engineers!" or, "Working to understand this phenomena is exactly what scientists do, so today we're *being* scientists." I believe this has a big impact on the way kids look to the future and sort out what they want to do with their lives.

- **When it's time to scale up, think spawning, not growth.** Small is beautiful in the world of tinkering. If you've got a dozen kids or more tinkering each day after school or each weekend, and you want to reach even more kids, you don't necessarily need them in the same place. What you may want instead are parallel tinkering spaces close to the other kids you want to reach.

- Finally, don't forget that **what you want to see happen is *thoughtful* tinkering**. You want to see kids thinking about what they're tinkering with. I assume that eventually, every kid will do this thinking, but it's good to make a conscious effort to get *them* to make a conscious effort to learn as they tinker. "Conscientization" was Paulo Freire's word for a different but related process. Here, what you want is for kids to become conscious of how they are learning, what there is to learn, and what they know already. Once they've got this ability, there is no stopping their self-education. Conversely, if they are not thinking, they are essentially back in preschool, led by the nose through a pleasant arts-and-crafts session, but not gaining nearly as much as they could from it.

Vignette: Jessica's Compact

Jessica, one of our fine high-school helpers, was greatly interested in cosmetics and decided to make a compact mirror that lights up when you open the cover. She made a tiny wooden box, cut a tiny piece of mirror, inserted a 9 V battery and a holiday light, and rigged a contact that completed the circuit when she opened the little cover. I suggested we show it on our monthly local cable program, and she said, with this brilliant success in her hand, "But where's the science in this?" As I pointed out the key concept of open and closed circuits, I reminded myself again that in the Workshop kids experience the science but are not necessarily cognizant of or able to articulate that science unless we discuss it with them.

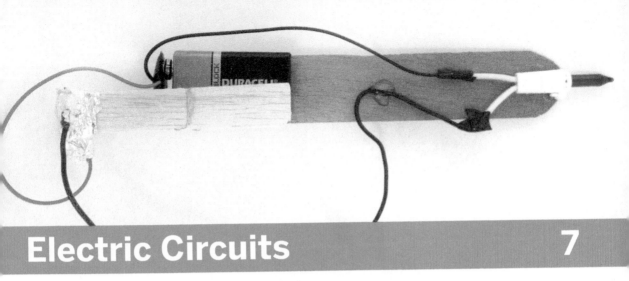

Electric Circuits — 7

Making a Flashlight, Magic Wand, and Steadiness Circuit

Tinkering with electricity is especially exciting because you can't see it, yet you know that it can flow through wires as well as the tissues of your very own body. Raymond F. Yates' 1942 classic, *A Boy and a Battery*, missed the mark by about 50 percent; it turns out kids love making lights light and buzzers buzz no matter what gender they are!

As I'm sure Edison would attest, were he with us today, making a lightbulb glow ranks among one of the deepest primordial sources of satisfaction. Perhaps the first people to make fire—and there were undoubtedly many firsts—experienced a bit more exhilaration, but a little shard of that original ecstasy can be seen in the eyes of every child who rigs a circuit to give light.

As you start your kids tinkering, here again is a brilliant opportunity to keep your mouth shut. Pass out batteries and lights, LEDs, buzzers, electromagnets, motors, and switches, then sit back and watch what happens, perhaps giving pointers here and there—but only as needed. This could take up an entire session with your group. No sense rushing for the final product until people have a feel for the phenomena.

You can do this free-form tinkering with snazzy little wood blocks, each mounted with a compo-

nent such as a battery or a motor, together with a bunch of loose alligator clip leads to connect them. Or, you can just strip a bit of wire on the leads of each of several items and have the kids twist them together in order to connect. *Figure 7-1* shows a hybrid arrangement with clip leads connecting a battery pack to an unmounted, stripped holiday light.

If you provide kids with just a few lights and a motor or two, the questions will flow, and the learning will take off effortlessly. More stuff adds more dimensions of complexity and more questions.

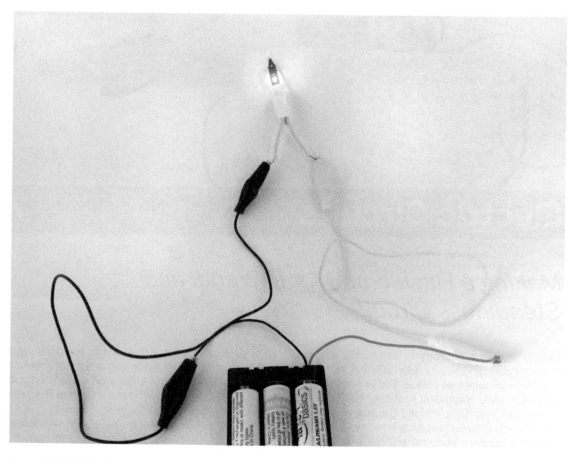

Figure 7-1. *Lighting the lights*

There's no need to go into any detail with explanations while they are tinkering. Just listen to and restate the students' questions, looking for clarity in the question itself. Once the circuits are working nicely, you can go back as a group and look for fundamental principles with which to describe them. You may come up with some like these:

- The batteries all have two sides, as do the lights (and buzzers, motors, etc.).
- Connecting to only one side of either the light or the battery is not good enough. All working circuits have at least one complete circle from one end of the battery to the other.
- The more lights are in a row, the dimmer they are.
- If several lights are in parallel, each one's brightness is nearly the same as if only one is hooked up.

You can have kids draw some of the circuits they create and then have them come to the board and draw them. Other groups can then follow these circuit models. This activity is well suited to presenting challenges like the following: Make a light glow and a motor go together! Or make two lights glow in two different ways! Or hook up two batteries in two different ways! Start with these entry-level projects and know that there really isn't any end. Your kids have started down the path known as *electrical engineering*, and only fate knows how far they'll go.

Make: Flashlight and Magic Wand

Gather Stuff

- PVC tube, 1/2" around and 10" long
- Holiday lightbulb (You'll want the small, pointy, cool ones shown in the photos here, not the old, bulbous, hot ones, which only light at high voltages. These small ones are still available widely at dollar stores and your local yard sale. The newer LED ones will also work fine and will demonstrate the principle of a diode, but it matters which direction the electricity flows, which may be an unwelcome complication with younger students.)

Be cautious about possible high levels of lead in the vinyl insulation. It is possible to find these sold with low lead content—IKEA offers them—but it is always best to have kids wash their hands after tinkering with anything vinyl or PVC. Finally, you can substitute small flashlight bulbs or LEDs if you want to avoid the holiday lights altogether.

- Three AA batteries
- Connection wire, around 22 gauge
- Paper clip
- Electrical tape
- Aluminum foil
- Paper

Figure 7-2. *The magic wand*

Gather Tools

- Wire strippers
- Scissors
- Hot glue gun and glue

Tinker

Step 1

Snip off a single holiday light and strip its leads. (Clip the plug off the end of the string of lights and throw it away to avoid someone trying to plug in the partial string.)

Step 2

Strip the ends of two connection wires a bit longer than the tube. Twist the stripped ends to the stripped leads of the light, one with a paper clip at the other end, the other with a bit of aluminum foil. Tape at least one of the twisted junctions so they don't touch and short the circuit.

Step 3

Put three AA batteries head to tail and roll them tightly in a piece of paper. Tape it to keep it from unrolling.

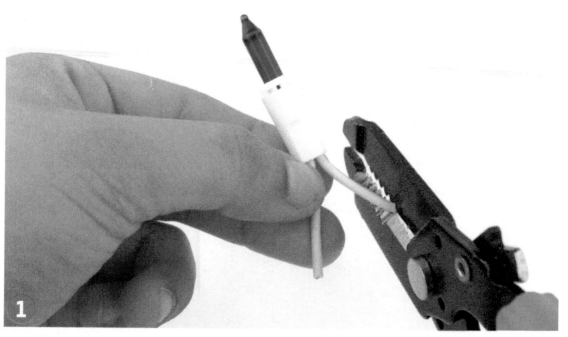

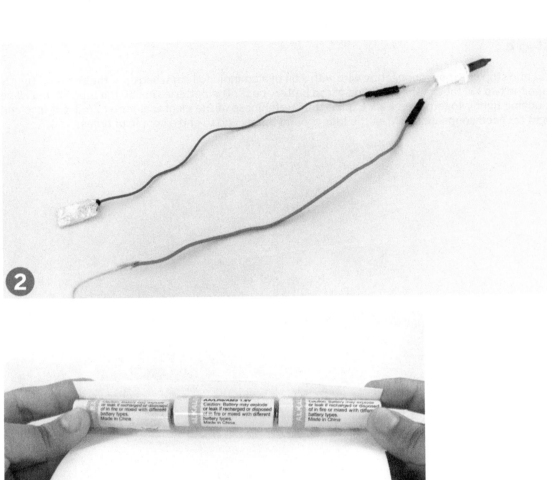

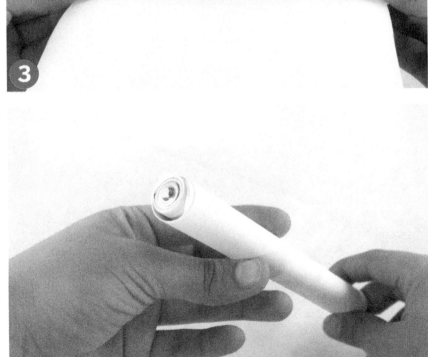

Step 4

Cut and strip one more connecting wire with a bit of aluminum foil smashed onto the end of it. Tightly tape the two foil tabs to the two ends of the battery pack. The batteries inside the tube should all be touching tightly together as well. Exploit the stretchiness of the electrical tape to make it nice and tight for good connections. Wrap the tape around tightly end to end a couple of times.

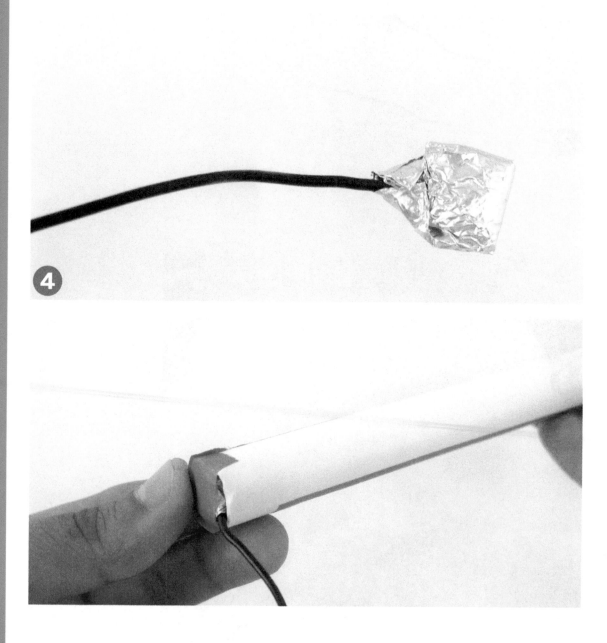

Step 5

Lay the circuit out just as it will slide into the magic wand: wire—battery pack —wire—lightbulb—wire.

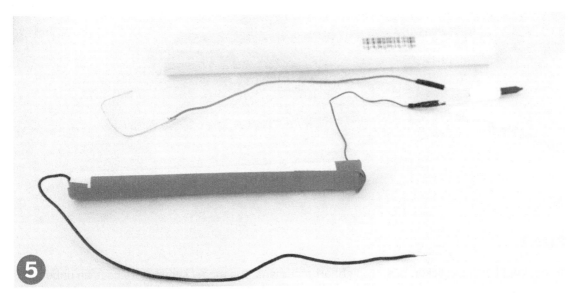

Step 6

Jam it all in the tube and make sure the lightbulb comes out the top. Now the two bare wires at the base can be connected in any way to make the light come on. You could just twist them together every time you want it to come on.

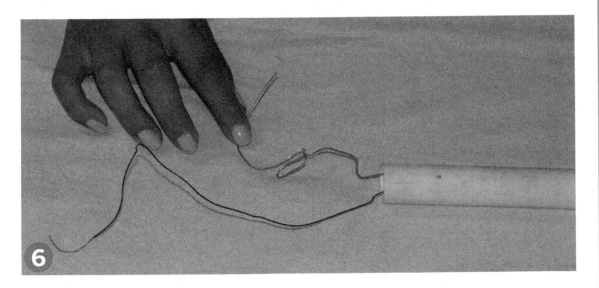

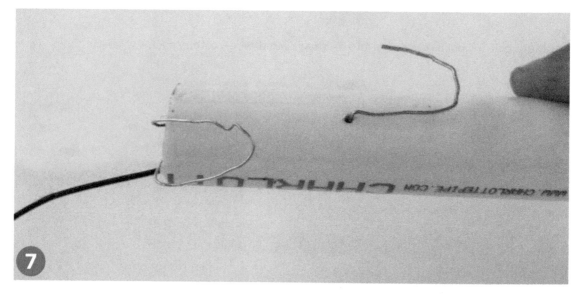

Step 7

Or you could make a slicker, push-to-light arrangement. The model shown here uses an unbent paper clip and a little hole drilled in the side of the PVC. (If you don't have a drill, you can melt that hole with a small nail held in a candle flame—ouch! Better get the pliers to hold it in the candle flame. And of course, make sure you have good ventilation if you go the melting route.) First the paper clip goes up through the hole from the inside out, with the other end hooked around the end of the PVC to secure it.

Step 8

The other wire then gets wound around the PVC so that when the paper clip is pressed up to the PVC, the bare wire and the paper clip contact each other. It takes some tweaking, but in the end it's nice because when you push, you're making the direct physical connection, which in turn makes the electrical connection.

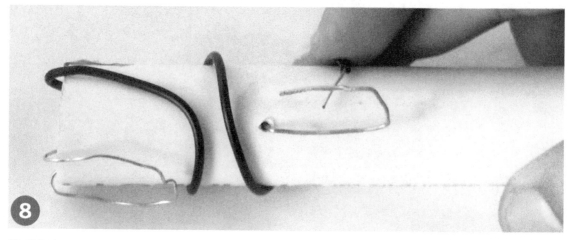

Step 9

Tape it all up (except for where you need the contact to be made) and glue or otherwise fasten the light into the top end of the wand.

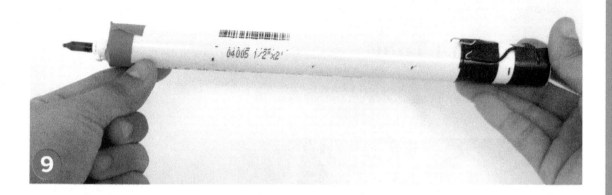

Check it out

- How do you reckon you could make the light brighter?
- How many volts do you have lighting your light? Those AA batteries are usually a volt and a half—check the label to be sure.
- If you check C and D batteries, you'll find they are usually also 1.5 V, but much fatter, heavier, and bulkier. Why would you ever want to use them when you've got these sleek little beauties?
- How do you figure you can make this light glow at all with these low-voltage batteries when it was made to be plugged into a high-voltage wall socket (120 V in the United States; 240 V in most of the rest of the world)?
- How is it that you're touching bare wire here but not getting shocked? Do you need to be careful with this project? (You can relax on that one: the answer is no.)

Make: Steadiness Circuit

- Base board—is a two-by-four is shown here
- Uninsulated, stiff wire
- Large nail, finishing nail if possible
- Large, wooden clothespin
- Tongue depressor or similar piece of thin wood
- Battery, 9 V with connector snap if possible (If you use LED lights, you'll only need 3 volts. Some holiday lights, LED as well as normal incandescent, can't handle the 9 volts, but those that can are gloriously bright, and that's what you want for this project.)
- Straw

Gather More Tools

- Hammer
- Side cutters (hefty), if you need to cut the nail head off

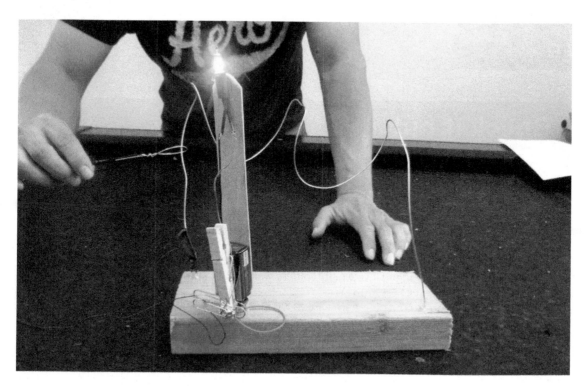

Figure 7-3. *Hold it steady and win the game!*

Tinker

Step 1

First make another flashlight, this one mounted on a tongue depressor with a clothespin as a switch. Hot glue a stripped holiday bulb so that it sticks off the end of the tongue depressor.

Step 2

Glue the 9 V battery onto the other end, with the leads sticking off. (If you don't have a battery snap, you can snip one off a broken device from your house, or just use paper clips to grab the little snap lugs, making sure to keep them apart so as not to short and drain your battery.)

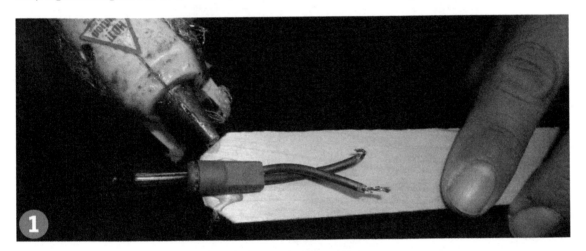

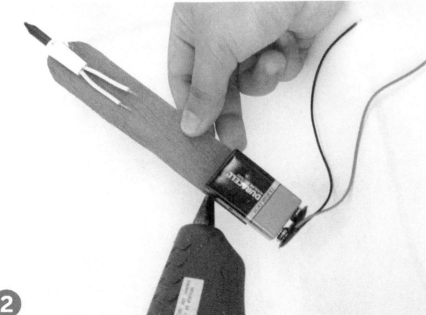

Step 3

Wrap little aluminum foil strips around the end of one of the wires from the 9 V battery and one additional connection wire. (Be sure to strip them first.) Fold the strips several layers thick, and make them around two inches long.

Step 4

Wrap these strips around the two handles of a clothespin. Put some glue under the strips so they don't fall off the clothespin, but keep the two surfaces where they'll come together clean to ensure a good connection.

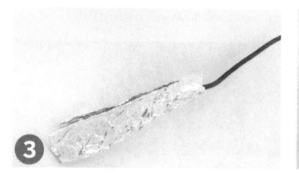

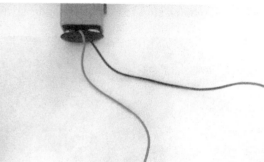

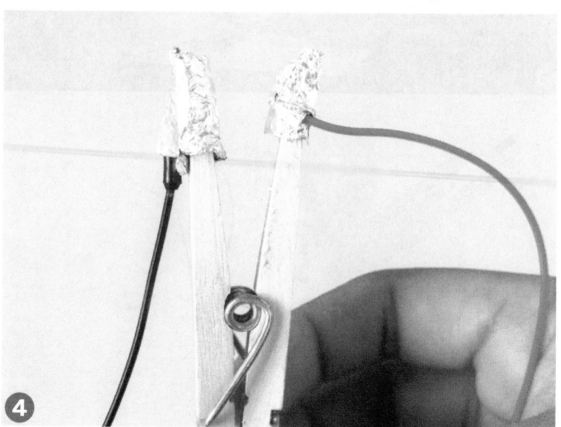

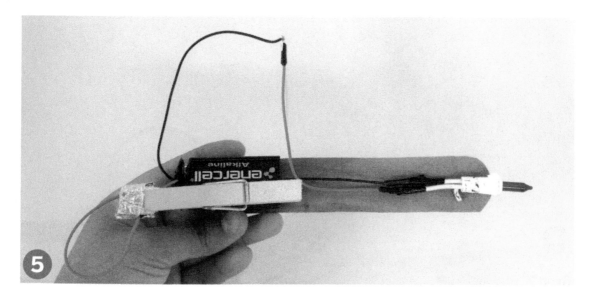

Step 5

Connect the other end of the loose wire to the remaining stripped wire of the lightbulb. Glue the clothespin onto the battery. It's going to work better to have the handles sticking off the end a bit. Now squeeze the handles together. If everything is in order, the electrons will flow, and the light will glow.

Step 6

Now make the base for the game. First bend the wand into shape by twisting a piece of stiff, uninsulated wire into a circle with a lollipop handle. The smaller the circle, the harder the game.

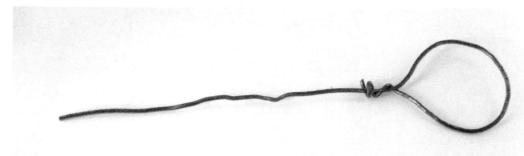

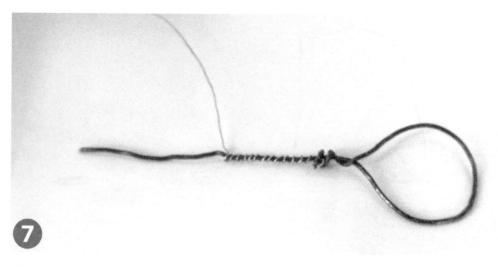

7

Step 7

Twist a connection wire onto this wire, making a good, tight connection.

Step 8

Drill or pound two holes into opposite sides of the base board. (You could use a nail for a drill bit for these small holes, since the wood is soft, and it is so easy to break small drill bits.)

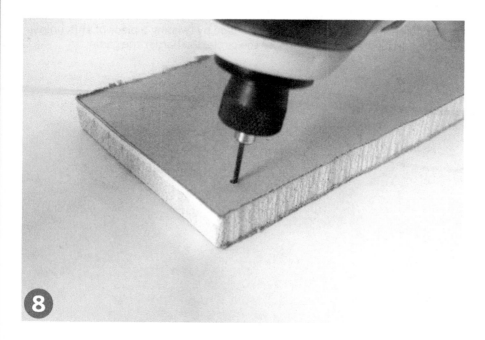

8

Step 9

Shape your course with a long piece of stiff wire, threading the loop-on-a-handle onto it and jabbing the ends into the holes in the base board. Glue them to make them stay. The bends may be altered later to make the game easier or harder.

Step 10

Get one last longish piece of connection wire, strip it, and twist it securely onto the long bendy wire.

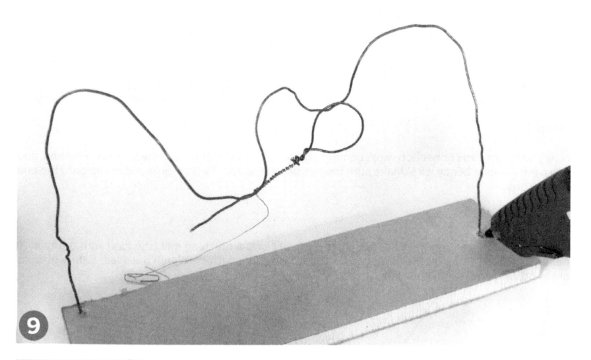

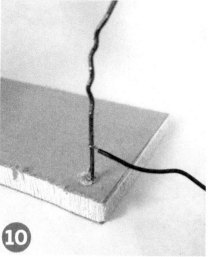

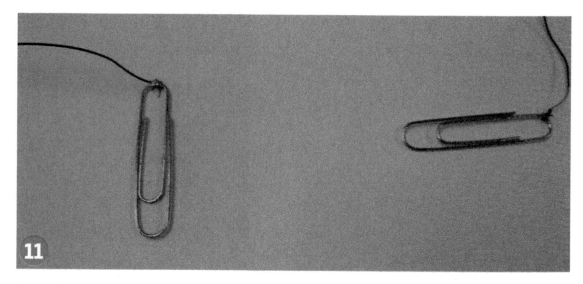

Step 11

Now you've got two connection wires coming off the base board: one from the loop-on-a-handle, and one from the long bendy wire. Make sure they've been stripped. Twist a large paper clip onto the end of each one.

Step 12

Hammer a large nail into the base board. If you can't find a finishing nail (the kind with a tiny little nubbin of a head), snip off the nail's head with the hefty side cutters. Watch the eyes! Safety glasses are good, or cover it with your other hand.

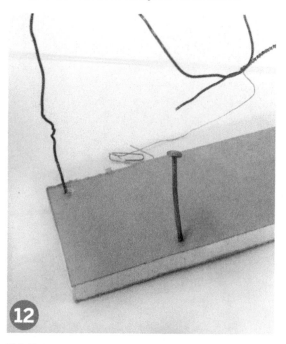

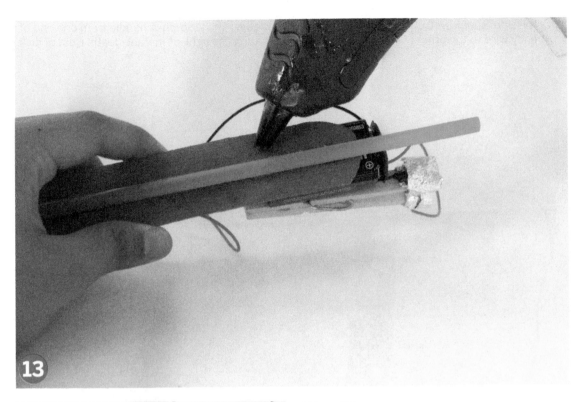

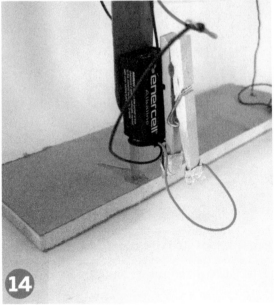

Step 13

Glue a straw onto the back of the flat flashlight. See in the image below how it sticks down farther than the handles of the clothespin? That's key.

Step 14

Slide that straw onto the nail you just hammered in, such that the flat flashlight now stands at attention beside the twisty wire. Clip the paper clips onto the two aluminum-foiled clothespin handles, and you're ready for action: the flat flashlight circuit now extends to the stiff bendy wire and the loop-on-a-handle.

Your mission, should you choose to accept it, is to maneuver the loop-on-a-handle from one end of the bendy wire to the other (*Figure 7-4*) without illuminating the light-of-instant-death. Best of luck.

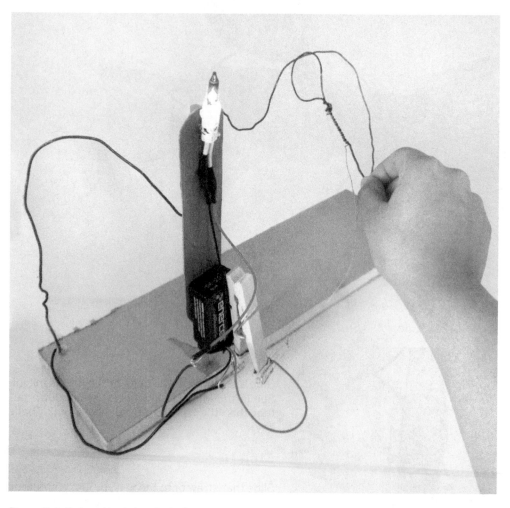

Figure 7-4. *Only a steady hand wins!*

 Notice the black tape around the stiff wire where it enters the base board back in *Figure 7-3*? This is a battery saver; you can leave the paper clips connected and drop the loop-on-a-handle there. The tape stops the connection. Unclipping a paper clip when you're not playing is also a battery saver. Of course, when you're done playing, you can remove the flat flashlight and use it for flashlighting.

 It seems like a bright idea to just connect up these happy little flashlights permanently, but the holiday light will sap the 9 volt battery quickly and the three AA batteries within about an hour, so it is best to keep them connected to the push-to-light switches: the clothespin in this project and the paper clip on bare wire in the previous project. This sort of switch is called a *momentary switch* in industry.

- If you wanted to cheat, how could you win the steadiness game every time?
- When the light is glowing, can you trace the circuit from one end of the battery to the other?
- Why do you suppose most flashlights don't have momentary switches?
- If you had a little buzzer and wanted to hook it up as well as the light, where would you connect it in?
- If you wanted the light to go on in the next room instead of on the same base board, how would you change things?

What's Going On?

Electric current flows through the wires in this project when the battery pushes on it and when the wires or other conductors present a complete circuit. Within the holiday light, there is a teeny-tiny wire that glows white hot when the current flows through it. That is the filament; it's an example of an incandescent light. (Fluorescent and solid-state (LED) lights don't use filament at all and are much more efficient, that is, they use less electricity while giving the same amount of light. But hey, we can handle a bit of inefficiency for the sake of a good, cheap tinkering project.) So this tinkering is all about making the circuit to let the current flow through that filament.

Did you notice that the switches in these projects are just breaks in the circuit? All switches are, no matter how fancy.

We used small batteries for this project, since the light is usually not going to be on for long. Bigger batteries may have the same voltage but they'll also have more energy waiting inside them. Electrochemical reactions provide the electricity in a battery, and when the reactants are used up, the battery is dead. Bigger batteries may have the same reactions giving the same voltage, but since they contain a lot more reactants, they have a greater total energy and last a lot longer.

Can you see examples in your life of larger batteries used for things that need more energy (motors and large speakers) and smaller ones for things that need less (LED lights, black and silver LCD displays)?

Usually those holiday lights are hooked up to the wall socket, which provides a much higher voltage. The catch is that they are not hooked up one by one, but rather in a string. When lights, or any resistors, are hooked up in a row to a certain voltage, they divide the voltage among themselves. Often there are 30 to 40 of those cute little bulbs hooked up in a row across the 120 volts of a US wall socket. That means they're getting 3 or 4 volts each. (The more there are, the less each gets.) This is called a *series circuit*. The downside is that when you remove one bulb, the circuit is broken, and all the rest go out.

Have you had frustrating holiday light experiences with series circuits? What happens when you take out a bulb from the string of holiday lights in your garage?

You can fiddle around taking individual bulbs out of a long string of holiday lights to see what the circuit really is. Often it is two or three different chains of bulbs in *parallel*. Thus, when you remove one bulb, that chain goes out but the others stay on. (When a bulb *burns out*, on the other hand, it may have a slick mechanism to allow the current to continue flowing through it so that the other bulbs remain lit. This was one of the great engineering achievements of the late 20th century.)

Can you find two or more bulbs in your string of holiday lights that have three wires connected to them, like in Figure 7-5? That's an indication that you've got more than one circuit in parallel.

To get 3 volts from AA batteries, you put two together in series. If you want 6 volts, put four together. If you rip apart a dead 9 volt battery, you'll find six little cells all hooked together in series. Go much above 9 volts, and your merry little bulb will flash and then turn dark and melancholy. There's only so much that little filament can handle.

You'll never get a shock from less than 12 volts, unless you put it across your tongue or similar body tissue. You can try pressing the connection snaps of a 9 V battery to your tongue and see if it's still charged. This is perfectly safe and very instructive: that little zing is electricity traveling through a wet tissue of your body. Your entire body conducts electricity, but the dry outer skin doesn't conduct very well, so it protects you a bit. Still, if you have enough voltage—120 is enough—and you lower your resistance by stepping into a puddle barefoot, you can die when the electricity flows through the wet tissue of your heart. Your heart is controlled by electrical impulses, and these can get fouled up when other electrical signals come through. If you're lucky, the paramedics will come quickly and bring a *de-*

fibrillator, which sends another electrical shock through your chest designed to set the heart signals clicking normally again. This has been shown to be even more critical than CPR for saving lives.

And, er, I hope it goes without saying that you should *definitely not* tinker with high voltages and your heart.

You win the steadiness game by not completing the circuit. If you want to cheat or impress your friends, secretly unhook the circuit in a different place, and the light will never glow! A buzzer would work in exactly the same position as the light, or you could try to hook them up in parallel, that is, hook in the buzzer without taking out the light. It could be that one or the other of them takes all the current. In a parallel circuit the electricity will look for the easiest path.

The circuit can be extended by longer wires. You can see that it could be used as a sort of security system: move the loop-on-a-handle just a hair, and the buzzer buzzes in the next room. Connect the loop-on-a-handle to the fridge door, and you'll know when someone is raiding the icebox. Gotcha!

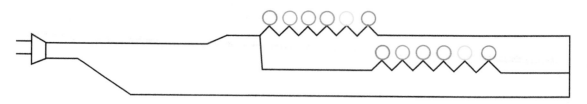

Figure 7-5. *Holiday lights diagram*

Keep On

You can make "glowies" (*Figure 7-6*) with a couple of tiny watch batteries and an LED. LEDs are directional, unlike the holiday light, so if they don't work the first time, switch the leads.

Security systems can be applied to nearly everything. A fifth-grade girl came to us with a plan to fortify her lunch box. *Figure 7-7* shows the result:

unless you disarm it with a secret lever, a terrifying wail is heard upon opening the lid. Safe sandwiches.

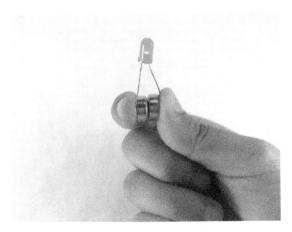

Figure 7-6. *A glowie*

Figure 7-8. *Wired for light*

Internet Conenctions

- Lightning is a massive current following a circuit through the air, cloud to cloud or cloud to ground. Search "lightning strike" on YouTube and try to figure how the lightning ended up striking in these places. Sure, it was the path of least resistance, but how so?
- Search "Faraday cage" and see if you can figure out why a car is a fabulous place to be in a lightning storm. Hint: The rubber wheels matter not.
- Want free energy? Get a solar cell, and you'll never have to buy another battery (until the sun goes down). Kelvin.com has got a lot of solar cells, kits, and gadgets.

Standards Topic Links

- Electricity, electronics, energy, circuits, series and parallel, components, conductors, and insulators

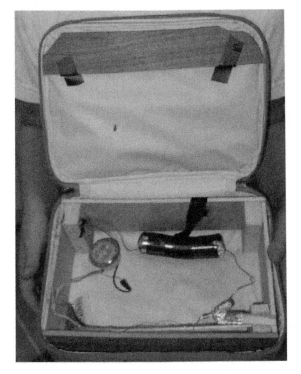

Figure 7-7. *Keeping sandwiches safe since 2013*

And naturally, if you go to the trouble to make a dollhouse, you'd best wire it up. You can see the light fixture on the top in *Figure 7-8*, but the light switch is just out of view under the mezzanine.

More Tinkering with Circuits

- Mike Rigsby, *Haywired: Pointless (Yet Awesome) Projects for the Electronically Inclined* (Chicago Review Press, 2009)
- Massimo Banzi and Michael Shiloh, *Getting Started with Arduino*, 3rd Edition (O'Reilly, 2014) Want a bit of higher-tech tinkering with circuits? Arduino allows you to tinker together a gadget and then control it with a computer. This book will jump-start you.
- Raymond F. Yates, *A Boy and a Battery* (a classic, albeit a bit dated, from 1942)

8 The Learning Community & Differential Learning

Tinkering is a social activity. Even loner geeks love the occasional group tinkering session. The communities of your students will overlap into the community you create around your students' tinkering. It's best to think this through a bit, informed by recent research, so that you can make your tinkering community a high-energy, high-interest, highly effective one. And there's one more thing you need to consider: your kids will all learn differently at different rates, but it's not difficult to support each of them just where they are.

The Learning Community

Tinkering often has a hard time fitting within the confines of the learning tradition. That is to say, if you want to teach well with tinkering, it may be best *not* to start with all the assumptions and culture inherent to the school classroom. Instead, you could consider the following question as a guide: how can I create the best little community for my tinkering kids? You'll need to get as much insight as you can into the broader community or communities you're dealing with, and within that context choose the best way to invite the kids' participation. The Community Science Workshops have formed deep roots in our communities, and each one has also formed a solid and well-grounded community of tinkerers who frequent our spaces. We want our work with the feeder communities to follow accepted cultural practices and to forward their own goals whenever possible. We pursue this by means of education through tinkering. Barbara Rogoff, senior professor of psychology at University of California Santa Cruz (coincidentally just up the road from the Watsonville Environmental Science Workshop), has researched and written extensively on different traditions of education used in different cultures and locations. Among others, she describes two traditions that contrast quite vividly and I think illuminate important differences between standard classroom learning and learning by tinkering. One is called *intent community participation* and the other *assembly-line instruction*.

Intent community participation, outlined by the "concept prism" shown in *Figure 8-1*, is what happens in many traditional cultures when kids learn by hanging around, listening, and participating in some productive activity. Kids can jump into the activity at various levels, and adults support them. Adults and kids and the entire social group gain from the experience. An example is kids learning basketry or agriculture or childcare by first watching, then helping, then taking on more and more responsibility as they grow older. This also happens in industrial societies in areas such as learning about home computers and learning religious or cultural activities.

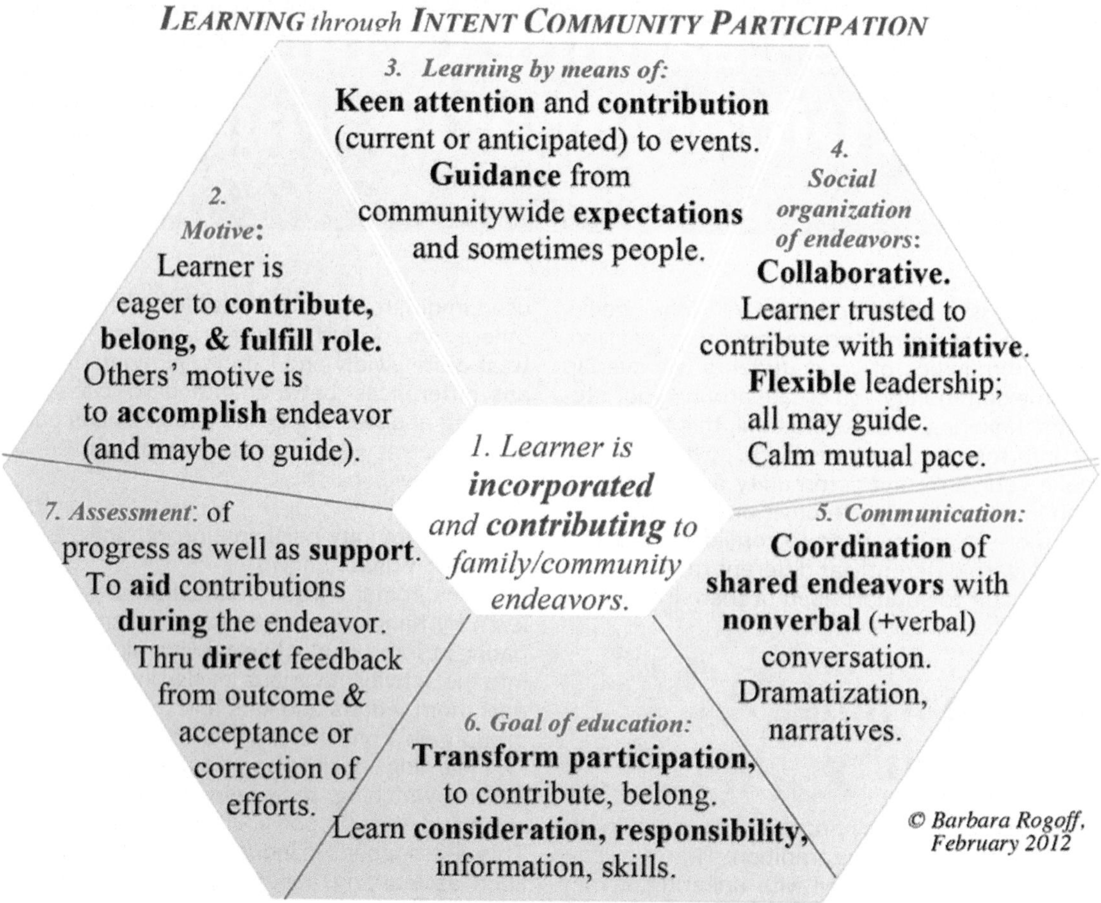

3. Learning by means of:
Keen attention and **contribution**
(current or anticipated) to events.
Guidance from
communitywide **expectations**
and sometimes people.

4.
Social
organization
of endeavors:
Collaborative.
Learner trusted to
contribute with **initiative**.
Flexible leadership;
all may guide.
Calm mutual pace.

2.
Motive:
Learner is
eager to **contribute,**
belong, & fulfill role.
Others' motive is
to **accomplish** endeavor
(and maybe to guide).

1. Learner is
incorporated
*and **contributing** to*
family/community
endeavors.

7. Assessment: of
progress as well as **support**.
To **aid** contributions
during the endeavor.
Thru **direct** feedback
from outcome &
acceptance or
correction of
efforts.

5. Communication:
Coordination of
shared endeavors with
nonverbal (+verbal)
conversation.
Dramatization,
narratives.

6. Goal of education:
Transform participation,
to contribute, belong.
Learn **consideration, responsibility,**
information, skills.

© *Barbara Rogoff,*
February 2012

Figure 8-1. *Intent community participation concept prism*

Assembly-line instruction, outlined by the conept prism in *Figure 8-2*, is what happens in many schools. Learning is separated from productive activity. Kids are separated from families and most other adults and have little agency to decide how and when they'll plug into the learning process. The content is decided by experts far removed from the community. Kids may have no idea why they're learning this content, nor any idea how to apply it, but the broader society has deemed it important. Students are expected to ingest the content and later they are sorted according to how accurately they can parrot back this content on exams.

Rogoff points out that some schools have moved away from assembly-line instruction, with great results in terms of their own goals. Clearly, different traditions developed to meet different needs, and these needs change. Rogoff and her group go on to draw various conclusions about children's development and learning of cultural practices and the broader consequences of different learning traditions. Their research can make you realize the value of thinking about how you were taught and your own assumptions about how education should be done, especially in comparison with other traditions. The key point for educators or, in our case, tinkering facilitators, is to know that you have a choice, beyond the nuts-and-bolts formats and structures mentioned in Chapter 4, as to how you invite your kids to participate in the tinkering you'll be carrying out. You'll make that choice consciously or not, so it's best to be aware of how you choose so kids can get the most out of the experience.

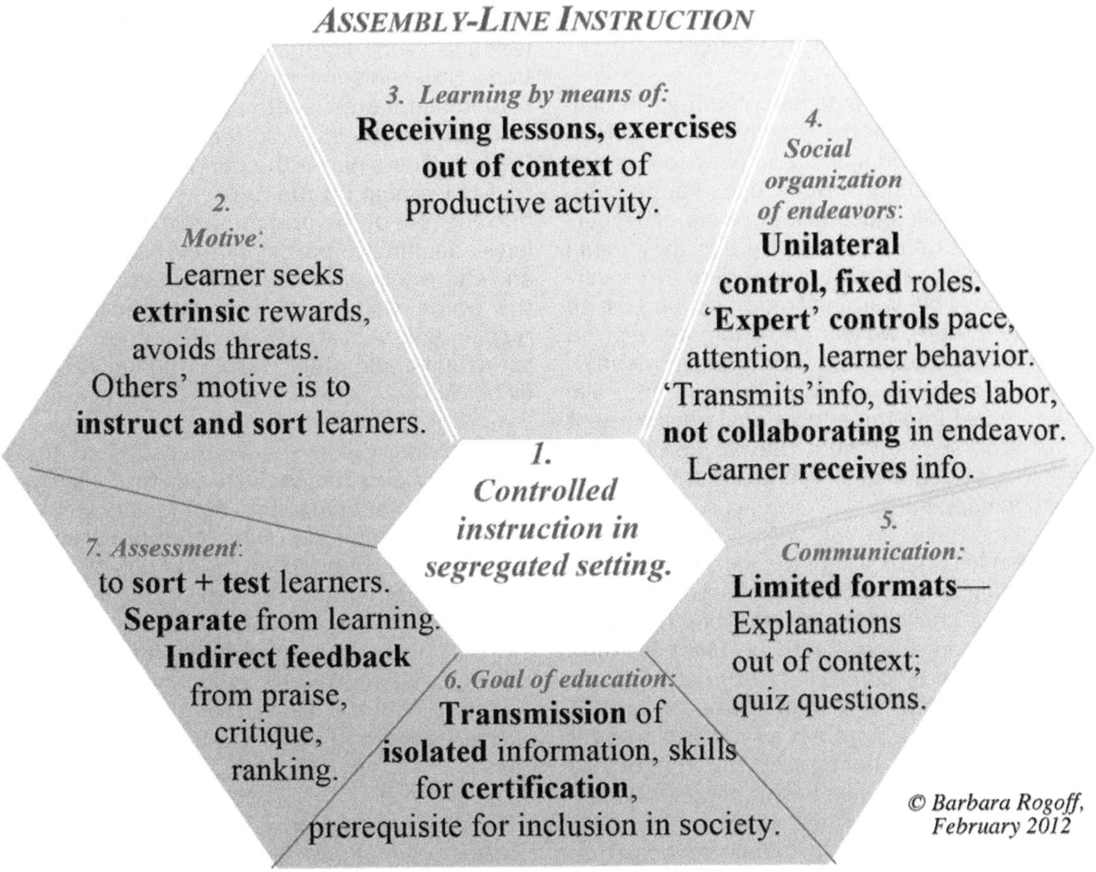

ASSEMBLY-LINE INSTRUCTION

3. *Learning by means of:*
Receiving lessons, exercises out of context of productive activity.

4. *Social organization of endeavors:*
Unilateral control, fixed roles. **'Expert' controls pace,** attention, learner behavior. 'Transmits' info, divides labor, **not collaborating** in endeavor. Learner **receives** info.

2. *Motive:*
Learner seeks **extrinsic** rewards, avoids threats. Others' motive is to **instruct and sort** learners.

1. *Controlled instruction in segregated setting.*

7. *Assessment:*
to **sort + test** learners. **Separate** from learning. **Indirect feedback** from praise, critique, ranking.

5. *Communication:*
Limited formats— Explanations out of context; quiz questions.

6. *Goal of education:*
Transmission of **isolated** information, skills for **certification**, prerequisite for inclusion in society.

© Barbara Rogoff, February 2012

Figure 8-2. *Assembly-line instruction concept prism*

Since it's hard to do good tinkering within the school setting, we have put some effort into forming our tinkering spaces at the Community Science Workshops more toward intent community participation. Some aspects are automatically in place. For example, there are usually many adults present in our tinkering space, some of them tinkering themselves. Kids of all ages are also present, and there is usually no time restraint, so the spectrum of watching/helping/doing often happens naturally. Newer learners want to contribute what they can, and more advanced tinkerers want to accomplish something. If assessment happens, it comes from direct feedback (Did the gizmo work or not?). Communication is about needed, practical issues having to do with the tinkering.

Other aspects are a bit harder to fit to the model. For example, productivity, per se, is not one of our top priorities. Sometimes we are productive, but it is most often the case that our tinkering goal is for fun, and that there are other, often easier ways to procure whatever it is we're making. Examples are constructing chairs, building toys, and sewing clothes. We also sometimes work together to offer something to the community, such as a parade entry or a haunted house. Here the goal is clear, everyone helps, and everyone benefits.

Rogoff says Intent Community Participation doesn't have to be oriented toward practical necessities or commodities, which is good, since I have concluded that the closest thing to an essential commodity we offer is an idea-rich, safe, educational space where kids can go and learn when they're out of school. That and the real learning that happens when kids are tinkering there. This learning happens nonstop amid all the fun, and I see that it mostly happens along the lines of Intent Community Participation.

Most kids raised in the dominant culture in California are weaned on the assembly line, school-based tradition, and it is not so easy to just pop in and out of traditions. But it does happen. Rogoff and her colleagues talk in terms of "repertoires of cultural practices." By this they mean that most of us have a bit of experience in several of the possible learning traditions, and as we grow and mature, we learn how to enter a given situation to participate in a productive manner. For many of our immigrant kids, it is naturally easy to enter into the educational environment of our Workshops. Where they've come from, there was a good deal of Intent Community Participation happening.

Education traditions aside, thinkers on teaching and learning long ago ceased to view education primarily as knowledge transfer, but rather as participation in a social activity. Most learning happens within a "community of practice" or a "community of learners," and it is nearly always "situated" within a certain physical and cultural space. Your kids will not be stepping from or into a vacuum of culture when they enter your tinkering space. As Rogoff writes, learning often happens by means of "transformative participation in shared socio-cultural endeavors." That means you'll see kids change and grow as they participate in your little community of tinkerers. Linda Polin similarly maintains that your kids will learn "through a process of enculturation into a slowly but constantly evolving practice." Your tinkerers will learn stuff as they become familiar with the norms of your tinkering environment, which is also constantly updating, if you will, in response to them.

Jean Lave and Etienne Wenger, early pioneers with these ideas, described it as actual "identity transformation within a community." They hatched a somewhat daunting terminology to describe what they saw happening: "situated learning" and "legitimate peripheral participation." So when some kid wants to learn to fix her bike, the first step is often to help someone else who is fixing theirs: hold the bike steady, fetch the wrenches, pump the pump, etc. This assistant is learning (in the future, she may step up into the role of the primary fixer) and also helping—a legitimate participant in the operation—and the whole kit-n-caboodle is situated within the warm, welcoming workshop environment, resplendent with tools, supplies, grease, interesting junk, and other people who know how to fix bikes.

These modern analyses from academia offer a whole new perspective for educators and open up a new set of guideposts that fit very well with tinkering. For instance, we go out of our way to support the community of kids in our Workshop to work together and see similarities in one another: We can all solder! We have all mastered the drill press! We all understand double-nut locking systems! We all know how to measure a quarter inch! We can all recite the safety rules of the Workshop! We're the Science Workshop Posse!

We also see our relationship with kids and their families to be of paramount importance. To understand their community and to help them acculturate into our little educational community is as important as any information we have to offer or how we offer it.

See Appendix B for selected references describing research done in this area over the last three decades. Check them out if you can, and don't get put off by the academic speak![1] The good news is that you needn't understand each nuance of these theories and frameworks. If you can just view your students as real actors in their own education, and view your program as a mini-community overlapping with the collective communities of your students, and see the par-

---

1.   Don't be surprised if you open up one of these references and find your eyes crossing at the site of what appears to be a completely separate language: "transformative enculturation through legitimate peripheral situated participation in a practice community..." It may inspire you to reject it all outright. But also don't forget the words of Kurt Lewin: "There's nothing more practical than a good theory!" And know that whatever happens in a given educational situation, it's following someone's theories and beliefs, conscious or not. Thus, it's better to think about it and make sure it makes sense. Then get back to the tinkering.

ticipation and interaction of your kids as essential to the learning that happens, you'll be primed and ready for wondrous education to take place with your group of tinkerers.

Differentiated Learning

In school, certain students tend always to be ahead of the game, always done with their worksheets early, always getting 110% on tests or looking a bit bored in class since they "know it all" already. Once upon a time, when I was young and right up until around 10 years ago, schools would skim off these students, put them together in a special group, tell them they have fantastic potential due to the talents and gifts bestowed upon them, and call this the accelerated group. There are huge problems with this, not least of which is the seed of arrogance sown in those who make the cut. And then there is the issue of the students who don't get into the winner's circle. Are they to be labeled the decelerated group? Slow? Untalented? Ungifted? No potential?

Some schools may still do this, but in California, no more. Now it's up to the teacher (of course with plenty of free time on her hands—not!—to cram one more thing into the instructional period) to carry out differentiated instruction. This means basically that the teacher should be meeting each student's current needs in relation to the content being delivered, with just enough challenge and just enough handholding to ensure productive learning takes place every minute of the day. This, with some students who have just arrived in the country, some students whose parents have been reading to them since before birth, students with all manner of disposition and personality and learning style and nutritional intake, all in the same classroom, often 35 or more of them. Sound impossible? Of course it is!

Teaching without real stuff is an awful challenge. Presenting knowledge to a group of students by means of worksheets or a textbook page is like feeding the inmates in prison: "Here's the grub, bub, take it or leave it." When teachers are pressed to carry out differentiated instruction, they're often given matrices or rubrics[2] full of activities or further exercises for the students to choose from, which can be a bit like giving the inmates two or three different forks to choose from.

When I've listened to differentiated learning zealots speak, yet another eating analogy comes to mind: intravenous (IV) feeding at the hospital. The bright hope is that the teacher will be able to discern and offer precisely what each student wants and needs in precisely the right manner using a stack of papers and workbooks, with maybe a video or two and, in the wealthy schools, the World Wide Web. An information IV, so to speak, of which the teacher is in charge.

Tinkering offers a better alternative. If the goal is to have students learn at their own levels and by means of the methods that work for them, why not let the students lead the way? And what better way to discern where to plug them in than to let them fool around with real stuff in real context? I'm not for a minute suggesting that pencils, papers, and reference texts should not be close at hand. I'm saying that these are not enough. It's a wide, wonderful world out there, but classroom walls and official curricula often cut it off from the students. In doing so, they also cut off many valuable learning avenues, some of which are essential.

Differentiated learning matrices will in no way substitute. Differentiated learning happens easily and naturally when students are tinkering. Students approach the stuff from their own standpoint, using everything they've learned up to now about this stuff and its place in the world. They also use tools and materials according to what they've seen and tried in the past. They are ready to learn the next level of knowledge this tinkering topic has to offer, and there are *always* more levels. I have seen the old adage proven again and again over a pile of interesting

2. An irrelevant yet irresistible aside: there were no rubrics when I was young. I'm convinced the rapid rise of rubrics, with their carefully crafted columns and categories of characterization is in large part due to the strikingly splendid sound of the spoken syllables in the word itself. Say it three times quickly. Rubric, Rubric, Rubric. Fun, eh? Like kissing a trout...

junk: the more you know, the more you know you don't know.

The beginning circuits tinkering activity outlined in Chapter 7 is a case in point. Some kids will need to learn that the insulation must be stripped from the wire to make it connect, while at the next table, other kids have formed a combination series/parallel circuit that shows a motor can let electricity flow without itself turning.

To do differentiated instruction, you simply approach students or groups *where they are* and help them move on. You can suggest a goal you think may be reachable and point the way forward.

In fact, doing tinkering without leaving it open to differentiated learning is darned-near impossible. I've seen it tried and it's painful and ugly: "Now pick up the hammer with your right hand. (Ana, put the glue down.) Now hold the nail with the left hand, three-quarters of an inch from the edge of the board. (Miguel, not that board.) Now give it a little tap to get it started. (Sofia, get back over here!)" And it gets worse from there. I guess that wouldn't actually be called tinkering, but perhaps "human robotics." Clearly the point has been missed: No one hammers the same. No one tinkers the same. No one thinks or learns the same. If we are to be facilitating our youth's learning, we've got to understand that and provide opportunities for them to do it in their own way. With real stuff, this is smooth and easy.

Beyond the actual act of tinkering itself, when it comes to observing and describing and trying to explain what's going on in the tinkering process, students also come at it from a plethora of different angles. The final discussion will bring this out, but you needn't wait. As they tinker away, your conversations with them can expose what they observe and what they do and don't understand. You are then well situated to offer them the support they need to get a conceptual grip and expand their understanding.

On a related note, *scaffolding* is another unfortunate term from the dictionary of edu-speak. Originally a scaffold was a structure on which the criminal stood to await his hanging. Hardly an appropriate goal for our schools. Scaffolding is just a funny word to describe individually supporting or helping a student to learn, and just like differ-

entiated instruction, scaffolding becomes utterly obvious in the tinkering environment. When the stuff is spread out, and the general direction is determined, kids start tinkering, and the astute teacher watches what's happening. Ah, Jill has never used wire strippers, and her buddy is not helping, so she'll need help from the teacher. Oscar is screwing off instead of focusing, which means he probably missed the directions or is afraid of not being able to build it. Melissa just cut way too much wire, so she probably needs a review of how to measure. And so on.

You can be as structured with this as you want. For example, just before launching the tinkering session, you can say you'll have a five-minute session on wire stripping at the front table for those who want it. You can ask for a show of hands on who understands, and then give more instruction if the majority of them want it. You can even put one or more students on a corollary project that is preparing them for doing their chosen project better, such as cutting a bunch of curvy lines with the scroll saw on scrap wood before they cut into the wood of their project.

Kids who've got the project figured out are valuable here, too, as mentioned earlier. You can either wait until they're done and then put them on assistant duty, or put the kids who've never done it before right with them from the start. It doesn't always work, but often it does. I remember one rough, annoying kid—a highly unlikely assistant that I ended up appointing to assistant on a snap decision (that is, I was about to snap)—telling me with a big smile at the end of the session, "You know, I think I like helping even more than building my own!" Score! I'd say he and the ones he helped were all well scaffolded and differentiated, so to speak.

(By the way, don't forget to learn from your students. It is a rare day when I do not learn something from my students, whether it be a new way to build something, a new idea on how to make something work, a new and plausible explanation for why something works, or a new mistake to make in the construction of an old project. When you give students the chance to tinker and ask them what they think is going on, you never know what you'll learn. Perhaps that could be called inverse differentiated instruction?)

Chemistry

Floating and Sinking with Colors, and Exploring Chemical Reactions

Ahhh, the wonders of a chemistry set! Mixing chemicals in a test tube and watching the resulting reaction is a visceral urge deeply seated in the psyche of every fourth grader. Safety is first, though, and many chemicals are not at all safe. Commercial chemistry sets try to profit from the urge while guarding the safety of the child. It is equally secure and quite a bit cheaper to limit the experimentation to common substances used for cooking (not the ones used for cleaning—that can get dangerous in a hurry). Even though there's no danger to speak of here, it's essential to pass out the safety glasses; it helps establish the habit as well as promoting lab fashion as chic.

Reactions, per se, are not required to do good chemistry. In the first activities here, there are only molecules sliding around each other. Solids, gases, and other liquids will interact with a base

liquid to the joy of all observing. Observing is especially important here; many subtleties may be overlooked without a keen eye to what's going on in the cup.

But reactions may be called the core of chemistry, so let's tinker with some of them, too! I'll show two nice ones that produce gas from the mixing of a solid and a liquid. The results are altogether different.

These tinkerings are all about observation and then hypothesizing the explanation. Unlike most of the tinkering activities we do at the Watsonville Environmental Science Workshop, with these we generally hold back the materials and hand them out one at a time to increase the chance that students will notice key phenomena, and decrease the chance that one student dumps the whole bottle into his experiment. Chemistry is bound to be messy, so round up some towels ahead of time. You can also use cafeteria-type trays to contain the mess.

Make: Floating and Sinking with Colors

Figure 9-1. *Experimenting with drops of color*

Gather Stuff

- Clear low/short cups (tumblers)
- Water
- Cooking oil
- Salt, the normal table kind
- Food coloring
- Stirring stick of some sort
- Some small objects to try floating and sinking:
  - Plastic beads
  - Corks
  - Beans or corn kernels
  - Rocks
  - Steel nuts
  - Real nuts
  - Toothpicks or bamboo skewers
- Other bits of stuff to drop into the cup
- Three transparent bottles
- Effervescent tummy tablet (Alka-Seltzer works well)
- Base board and mounting system for horizontal bottle

Gather Tools

- Knife or box cutter
- Magnifying glass

Tinker

Step 1

Pour a finger's thickness of oil into the cup.

Step 2

Pour in water until their is a good amount of liquid in the cup. Notice the position of the oil and water.

Step 3

Dump salt into the cup, a burst at a time, and watch the action through the side of the cup. Try to figure out what's going on there, up and down, taking into consideration the four substances sliding back and forth: air, water, oil, and salt.

Step 4

Drop a couple of sizable blobs of food coloring into the cup. Watch carefully, but don't stir, if you can resist it. This may take a couple of minutes. Whip out that magnifying glass if you've got it.

Step 5

When you're done watching the miniature, inverse eruptions, stir it up good and then observe it as it settles down. You'll notice some more amazing stuff.

Step 6

You can continue using that cup full of stuff, or pour yourself a fresh one of water and oil, with no salt this time to make the view clearer. Assemble a bunch of objects to toss into the cup.

Step 7

Drop them in, one by one. You can try to predict where they'll sink to before you drop them. Keep dropping stuff in until you've got things resting on top, on the very bottom, and at the boundary between the oil and water. It's a bit hard to catch this on camera, but in the second image, you can see a toothpick and a piece of cork floating on top, plastic beads floating in the middle, and beans and a steel nut resting on the bottom. Again, you can stir it up and see what happens.

Step 8

There are several other cool things to tinker with on this same topic. Here, the cup on the right has salt water at the bottom, no food coloring, and fresh red-colored water at the top. We used the straw to gently move the red water to the top of the salt water—cover the top of the straw with a thumb and move a couple of inches of red water over and let it slowly out on the surface. If you just pour it in, they tend to mix. You can try for three layers if you make one cup of super-salty water, one of medium salty water, and one of fresh water. When you're done observing, stir it up and see what happens.

Step 9

So far, we've tinkered with liquids and solids. Here is an arrangement in which you can introduce gas into the mix. It's a bottle mostly full of oil, with only an inch or so of blue-colored water at the bottom. In goes the effervescent tablet, and, wow, watch what happens. (You **must** leave that lid off for safety; wouldn't want to build up any pressure in there.)

Step 10

Put a flashlight underneath and flick off the lights: party time!

Finally, you can fill a transparent bottle completely with oil and water, slightly more oil than water, then build a tiny boat out of material that floats between the water and oil. Build a fancy mount for it, and you'll have an executive desk toy worth $40 in the designer catalogs (*Figure 9-2*).

Figure 9-2. *Your desk toy*

Since it's a bit hard to see the boat within the bottle, *Figure 9-3* shows it separate. We used plastic from a milk bottle, a section of bamboo skewer, and a bit of plastic bag for the flag.

Figure 9-3. *A close up of the boat*

Check it out

- What was the salt doing as it sunk through the oil and water?
- What was it that came back up sometimes after the salt hit the bottom, and why?
- Why do you think that food coloring wasn't keen to color the oil?
- Why do you think it sunk all the way through the oil eventually?
- How does a drop of food coloring spread itself out in the water once it hits?
- What determines how far a given thing—solid or liquid—sinks in the cup of oil and water?
- How do the gas bubbles from the tummy tablet interact differently than the solids and liquids?
- How does adding salt to water change where it floats?
- What determines how that tiny boat floats, since the wood mast, the plastic base, and the plastic bag are all different materials?

Make: Chemical Reactions

These two activities require the use of a match or burning splint. Attempt these only with adult supervision, and be sure to have a fire extinguisher handy.

Gather More Stuff

- Baking soda or baking powder
- Vinegar
- Peroxide
- Yeast
- Matches
- Craft sticks
- Cork, to fit bottle
- Napkin

Tinker

Step 1

This must be the granddaddy of all chemistry activities: mix baking soda and vinegar. Here we do it in a cup, since we'll want to analyze the gas that is produced. The exact proportions are not important; make it fizz, baby.

Step 2

After it's fizzed a bit, light a match or a wood splint, and lower it into the cup. What happens?

Step 3

Now for the second reaction: hydrogen peroxide and yeast. Mix'er up.

3

Step 4

Whip out the matches or the wood splint again. Get one burning, blow it out, and lower the glowing tip into the cup. Observe.

4

In each of these activities, you can also feel the sides of the cup to see if they get hot or cold.

I can't resist showing you one more that I got from an old science activity book when I was a kid. It's a classic's classic: shooting a cork with the pressure from a baking soda and vinegar reaction. First get a bottle and find a cork that fits it snugly. Put a bit of vinegar in the bottle and some soda in a napkin (*Figure 9-4*).

Figure 9-4. *Prepping your ingredients*

Go outside. Roll the napkin up around the soda so that you can stuff it into the bottle (*Figure 9-5*).

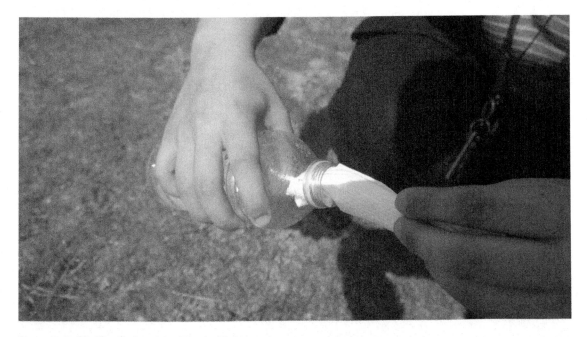

Figure 9-5. *Stuffing the napkin in the bottle*

Figure 9-6. *Cork it, and get away!*

Keep your face away from the cork once you've placed it! It's always best to wear safety glasses.

Contemplate the universe as you wait for chemistry to happen (*Figure 9-7*). And then physics.

Figure 9-7. *Ready to pop*

- What happens when you change the amounts of each of the ingredients above? (To be real scientific, you should only change one at a time; that way you can see the influence of each factor.)
- Why do we use fire here to check out the gas produced?
- Since nothing more is added to the bottle after the cork is placed in the final activity, where does the pressure come from?

What's Going On?

The tinkering activities on floating and sinking here show how far you can go just putting different materials together and observing them. Even sticking to totally safe kitchen materials, this is the tip of the chemical iceberg. Think of the other stuff you could mix together, and see what happens: different oils (olive oil or body oil), beer or stronger alcohol, fizzy drinks, soap, milk, sugar, ice, honey, peanut butter, chocolate, mustard, Jello...[1] Each of these things will interact with the others in a specific manner linked to its chemical properties.

One of the main observations we make while tinkering with this stuff is which item goes up and which item sinks. This is a pretty big question. If you throw a tiny pebble in a lake, it sinks, but an entire tree uprooted by a hurricane and carried out to sea floats. So do giant oil tankers and cruise ships and perhaps the biggest things ever to float on water, enormous icebergs that have cracked off an ice shelf on Antarctica. Naturally, these things are all heavier than the pebble, so what's up?

Floating and sinking are often trivialized by the following (accurate) statement: things of lower density float atop things of higher density. This only answers the question if you understand density and gravity.

Density is how much mass there is in a given volume. Take two identical bottles— same volume— but fill one with air and the other with sand. Cap them tightly, chuck them in the swimming pool, and you'll see that you can change density just by changing mass and leaving volume constant. Now take a little rock and hack out a piece of Styrofoam so that it has exactly the same mass as the rock. (You can do this on a balance made from a ruler sitting crossways and centered on a pencil.) The Styrofoam will be much bigger— that is, it'll have a bigger volume. Chuck these two in the pool and you'll see that density can be different even if mass is constant, as long as the volume is different.

Technically speaking, density is calculated by measuring the mass and the volume of something and then dividing the mass by the volume. Density = mass/volume, in the language of mathematics.

The density of the water in those examples above is key. The things that sink have a greater mass than an equal volume of water, that is, they're more dense than water. Things that float are less dense. This is true in all fluids, whether the floaty/sinky things are solids, liquids, or gases.

A fluid is something that flows; basically it's something that's not solid. Water and air are

---

1. Wait a minute, what the heck *is* Jello? Animal, vegetable, or mineral? My childhood buddy Brad said it was his favorite food. What did that mean for Brad's metabolism?

both fluids, and things float or sink in both water and air.

You may say, well, things have to be pretty darn light to float in air! To which I'd reply, yes, lighter than air, or more accurately, using our new understanding of density, they have to be less dense than air. They have to have less mass than an equal volume of air. Hot air and helium balloons, as well as the air surrounding a fire, all meet this condition and float, that is, rise in air. They feel an upward force.

To understand where the force comes from that pushes up on something that's floating, you have to look the other way: gravity. Say you've got bottle full of air, tightly capped, and you're pushing it under the water. It wants to float, but you're forcing it under. The force you feel pushing up is called the *buoyant force*. Here's where it comes from: the bottle is taking up space in the water. If you let the bottle go, it will rise, and water will rush in to take up the place where it used to be. That water rushing in is pulled by the earth's gravity, which always pulls things down. So basically the bottle is pushed up by the water being pulled down.

The bottle is being pulled down, too, mind you, but with less force than the water, because it is less dense. So the water wins and the bottle floats.

Thinking about a discrete bottle is simple. Thinking about an oil slick is a bit more complicated. Every drop of the oil is less dense than water, so every drop will float on top of the water. Likewise, in the first activity, the food coloring is more dense than the oil, so any little drop will eventually make its way down to the bottom.

Think about all the things you've seen float and sink. Can you understand them in terms of this explanation? When you are floating in a pool, do you see that your density is less than the water?

Now that I've got that density/gravity/floating/sinking issue off my chest, let's look at each observation. Salt going into the cup with oil and water goes to the bottom since it's more dense. On the way down it gets covered with oil, but takes some air with it. This air eventually burbles away from the salt and glorps back to the top, maybe

with a bit of oil. This is because it's less dense.

Did you notice a bit of oil and maybe air staying with the salt at the bottom? As long as the whole blob together has a density greater than water, it will stay down there.

Common food coloring is made to color water-based foods, so it doesn't mix with oil. Once it makes its way to the water, it mixes quite well, following previously invisible motion in the water. Some of this motion is currents left over from you pouring the stuff in and jostling the cup, and some of the motion is innate to the fluid molecules, and is called *Brownian motion*. Essentially all those water molecules are knocking into one another, and when the food coloring comes in, you can watch it get knocked into as well. Eventually it spreads through the entire volume of water.

You can get oil-based food coloring at fancy food shops and on the Web. It's used to color butter and other greasy things. If you can get a hold of some, tinker around with it in the same way to see what happens.

When you stir it up, the densities of these things don't change, and they don't mix much, so they return to their original spot. If you dump soap into the cup, there will be some mixing as the soap molecules pull on the water molecules with one hand and the oil molecules with the other. In the activity with salt water and fresh water, they'll mix together and never come back apart. This makes sense, since the salt is ready to mix into the fresh water just as it did to make the salt water. Oil and water are different enough that they won't do that.

With the solid objects, again it is simply a question of density and gravity. Those sitting contentedly in the center have densities greater than oil but less than water. The parts of the little boat in the bottle are all connected, so its combined mass and its combined volume determine its density. At the same time, the mast may be a bit less dense, tending to keep it above the hull—convenient, since that's the way boats should be, even if they're at the bottom of a sea of oil.

Adding a solute to a solvent, that is, dissolving something in a liquid, tends to make it more dense. Thus, we get the layered salt water. Tummy tablets release gas upon contact with water. This gas is much, much less dense than the water or oil, so it takes off upward with vigor. It may bring a bit of colored water with it, much as the salt brought some of the oil down in the first activity.

Now on to the reactions. Hydrogen peroxide can be viewed as water with an added oxygen: H_2O_2. In Spanish and other languages, it's spelled out clearer—it's called "agua oxigenada," oxygenated water. Yeast is little specs of dormant fungus, and when you mix it with the peroxide, this is enough to set free that additional oxygen and leave yeasty H_2O in the cup. Other things work also; you could try adding salt to release the oxygen from peroxide.

Baking soda is sodium bicarbonate ($NaHCO_3$), and vinegar is watered-down acetic acid (CH_3CO_2H). Baking powder is mostly baking soda together with other stuff that tweaks the reactions in baking. When those two kitchen chemicals come together you have a base reacting with an acid with a resultant of carbon dioxide.[2]

Both oxygen and carbon dioxide are gases at room temperature, so in both cases you have the thought-provoking situation of a solid (soda or yeast) and a liquid (peroxide or vinegar) coming together to create a gas. The properties of these gases are quite different. One promotes burning; oxygen is the key element in many combustion reactions. The other stifles burning; some fire extinguishers are filled with CO_2. This can be seen nicely with the burning splints.

Have you ever fanned a fire, or blown on it to get it going? You were force-feeding it oxygen, just like you did when you lowered the glowing splint into the peroxide and yeast cup.

These gases are formed in the bottom of the cup and tend to stay there. Again, the issue is density, but since molecules all take up the same volume when they are gases, we can just compare the masses. The mass of air can be estimated by taking its largest two components: nitrogen at 78% and oxygen at 21%. (The last percentage point is argon, carbon dioxide, and a bunch of other stuff, and then there is water vapor, which can reach 3%, but let's ignore these for now.) Nitrogen and oxygen both tend to bond with themselves in pairs. Nitrogen has an atomic weight of 14[3] and oxygen 16. Doubling those numbers for the mass, the weighted average comes to $0.78(28) + 0.21(32) = 21.84 + 6.72 = 28.56$.

Carbon dioxide, CO_2, has a mass of 44, and oxygen, O_2, 32. Both of these are more massive, that is, denser than air. So both the carbon dioxide produced in the baking soda and vinegar reaction and the oxygen liberated in the peroxide and yeast reaction tend to stay sunk in the bottom of the cups, allowing a bit of time for analysis by dunking in burning splints.

The flying cork activity shows that when a gas is formed in a closed space, it can create a high pressure. When molecules turn from liquid or solid to gas, they take up 1,000 times more space, so it makes sense that the more gas is formed in the bottle, the higher the pressure gets until... kabloom.

Did you have the experience of making a bottle of vinegar and baking soda that would not blow its cork? It takes a certain amount of pressure to blow it, and if the reaction is over and no more gas is being created, the cork will never blow. Solution: Next time add more reactants!

2. As usual, it's a bit more complicated. Vinegar has got a lot of water in it, and the acid takes the form of disassociated ions in the water. Baking soda also disassociates into ions when it hits the water and vinegar. Two of those ions react to create carbonic acid (H_2CO_3), which is unstable and deteriorates straightaway to give water (H_2O) and carbon dioxide (CO_2).

3. Fourteen what? Fourteen grams per mole of atoms. What's a mole? A whole bunch of something, 6.022×10^{23} to be exact, and that's a monster of a big number; around six hundred thousand billion billions.

Keep On

As long as you've got the baking soda and vinegar mess out, you may as well make the classic volcano. This one (*Figure 9-8*) was made with plaster of Paris in the neck of a two-liter bottle, with a hole drilled for the reactants. Soap and red food coloring go a long way toward making a realistic volcano. We pour the vinegar in last, and then out comes the lava to flow down the side of the volcano (*Figure 9-9*).

Crystals are another fabulous and safe area to tinker in. *Figure 9-10* shows some crystal jewels made on pipe cleaners dunked into a super-saturated solution of borax and a bit of food coloring.

Last, your chemistry tinkering will not be complete without making gak, also known as slime. White glue, water, and borax, with or without color, will give your kids hours and hours of polymer delight. Change the proportions, and you'll get different gak! *Figure 9-11* shows it making its way down the sides of an overturned cup.

Figure 9-9. *Foooosh!*

Figure 9-10. *Crystal jewels*

Figure 9-8. *Loading the volcano*

Figure 9-11. *Gak*

Internet Connections

- Search "electrolysis" on YouTube to see water molecules ripped apart by an electric current. You can do it with a few batteries and a cup of salt water.
- Search "sulfuric acid sugar snake" to see a somewhat more dangerous experiment between acid and sugar.
- All crystals are formed by materials solidifying out of a fluid. Check out photos of the world's largest gems and geodes (crystals inside of hollow rocks).

Standards Topic Links

- Density, mass, atomic weight, chemical reactions, acids and bases, states of matter, floating and sinking, and buoyancy

More Tinkering with Chemistry

- Simon Quellen Field, *Why Is Milk White?* (Chicago Review Press, 2013) and *Culinary Reactions: The Everyday Chemistry of Cooking* (Chicago Review Press, 2011)
- Virginia Mullin, *Chemistry Experiments for Children* (Dover, 1968; a classic from the 1960s)
- Pat Murphy, *Boom! Splat! Kablooey!: Safe Science That's a Real Blast* (Klutz, 2009)

10 Dealing With Questions and Dishing Out Answers

Tinkering generates questions, and as the facilitator, you'll be looked to for answers. Relax: you don't need all the answers, and in fact, it is much more important to have the right approach to dealing with questions than an inventory of all the answers you may be called on to provide.

Questions

Questions are the basis of learning. It seems to me extremely difficult to learn something meaningful without having a question about it first.[1] Tinkering is valuable in the learning process precisely because it raises real, relevant questions.

If you're a teacher like me, more than once you've been in a dreary classroom trying to convey some content to your charges and failing magnificently to make it come alive for them, when suddenly, in response to some unseen cosmic kick, one of them scratches his head, puts up his hand, and asks a pointed question. The room comes alive. Other students join in with tentative answers and additional questions, challenges are offered and countered, and education takes off out of the rut, pedal to the metal, accelerating into the wild blue yonder.

I even said to myself a few times after such occasions in my early teaching career, "Is there any way I could plant a question like that? Maybe bribe a student ahead of time?" I thought it surely possible to make this happen on demand; it's so good!

It turns out that tinkering is a quite reliable way to raise those questions. If, for instance, you tell a student that force equals mass times acceleration, she is left with few choices as to her response: accept it as truth, reject it as a lie, or ask you to explain or prove it. If, on the other hand, she has just designed and built a catapult, the questions will rain down like a hail of arrows from the castle turrets: Will a heavier or lighter rock go farther? Will a change in elastic make it go farther? What is the best launch angle? Is there an optimal force, or is more always better? Can you correlate force with distance? What is the limit to scaling up a siege engine? What was used as elastic a thousand years ago? Is there a way to power it with just gravity? What are the key factors to making it throw far? Why does that group's catapult work better than ours?

These are all questions that will set students to learning. Each one can be explored, and some answers can be arrived at using nothing more than the materials on hand. At the same time, answers are not important just yet.

The value of the student's question is supreme. The best initial response to a question is not to answer it, per se, but to validate it, protect it, support it, and make a space for it. Like a blossom just emerging, a question is vulnerable and delicate. A direct answer can extinguish a question if you're not careful. But if you nourish the blossom, it will grow and give fruit in the form of insight as well as more questions.

In short, a question needs to be nurtured more than answered. It should be given center stage, admired, relished, embraced, and sustained.

---

1. You may protest by saying that you can learn many unexpected and fascinating things just by keeping your eyes open and staying mindful of the surrounding universe. To this I reply that your mind is thus open with an ever-ready, curious question on anything. This is a truly sublime way to live, and perhaps a paradigm for lifelong learning. This supports my thesis here.

Practically speaking, here are some strategies my staff and I use to nurture questions that arise in our Workshop:

- Commend and praise the question. I'm not a believer in praising the student, but to praise a question is to show others what is valuable. Don't be afraid to shout it out. In fact the answer that is *always* right, *always* good, and *always* ready is: "Yeah!" And then repeat the question with a huge smile.

  Student: "Why does the sound from the drum change when I hit it with a different stick?"

  You: "Yeah! (grinning) Why *does* it change with different sticks? Excellent question; what do you think?"

- Restate the question. There are no stupid questions, but there are "wrong" questions. By this I mean that what students really want to ask is not what they are asking. If you restate and slightly realign the question, they may be even more pleased with it.

  Student: "Why does the sound get higher as the drum's head gets stiffer?"

  You: "Yeah! Why does the pitch of the sound go up when you tighten the drum head?"

- If a certain question arises that is key to getting at an understanding of the concept, or if a question has really gripped a student or group, write it on the board. You can even collect all questions on the board after spending a certain time tinkering. This demonstrates to all the value of forming questions, and everyone gets to benefit from the questions of others.

One category of especially valuable questions is comprised of those that, with a smidgen of guidance, allow your students to definitively work out their answers on the spot, in the course of a few minutes, with the materials and instruments on hand. When this sort of question pops up, it's all I can do to keep from salivating and telling them, like some kind of pusher in a dark alley: "You,

lucky ones, are about to learn something from your question, straight from the stuff on this table that you've been tinkering around with, and in doing so you will largely bypass myself and all other authorities. You will make a dynamic and dramatic learning advance that will serve as a model for infinite future possibilities in your lifelong quest to understand! Hold on, here it comes!"

Know that the best questions, the stimulating and driving questions, the roots of learning questions, are not *your* questions but rather the students'. Many teachers have recognized the value of questions, and have replaced their lectures with a series of questions. Unfortunately, this can have the dour effect of turning the class into a never-ending quiz. Many students are heavily conditioned from years of practice to try and please the teacher by thrashing and grabbing for the "right" answer, so even the best of the teachers' questions can easily turn into a trivial guessing game. Sometimes even when I'm rephrasing and restating a student's question, she'll take a wild stab at the answer, thus trivializing the moment and snipping off the tender shoot. This is a real danger.

When we have questions that we hope the students will make their own, rich questions that we know will lead to deeper understanding, we have to be clever. Just as with the toddler who will embrace his own decision and reject yours, it is often best to arrange the situation so that the students stumble right into your question. The more times you can repeat a tinkering activity, the better you'll get at setting these delectable traps for your students.

This is in no way underhanded, and you needn't be dishonest in your response. Good gurus have been doing this for millennia. For example, if you make the model of a simple circuit such that two of the bare wires are likely to cross, when students begin constructing their own circuits, some will inevitably cross the wires, leading to a short circuit, thus preventing the desired function of the set of components. "What's going on?" ask the fortunate ones who encounter this situation. Gotcha. "Yeah, well let's see. Try following the circuit with your finger, and see if you can find the problem."

When good questions arise that the students are excited about, it's an indication that good learning is happening. Minds are engaged, focused, and open for more understanding. Questions give you a view into what the student understands and does not, a form of assessment. I'd much rather have a couple of questions by the student than completed quiz papers. More on that next.[2]

Answers

Giving answers is a tricky business. As you develop them and hand them out, think like an artisan: try to craft the very best ones possible. There is a huge difference between good and mediocre answers.

In the previous section, I've tried to make the point that to naively respond to a question with an answer full of info is often deleterious to real learning. Questions, like blossoms, should be nurtured above all. The question is infinitely more important than the answer. Think about it: answers are useless without questions. Here's an answer, for example: 3.14159. Another: the Coriolis effect. Clearly, these answers have no meaning at all when orphaned from their questions.

I may be pushing it here, but I'll go so far as to say that answers are cheap. Textbooks are full of them, as is the Internet. The questions that motivated finding those answers are the gemstones, the real heroes in the story. Furthermore, only with the question can one determine which answers are right.

Nevertheless, when the student has articulated a solid, well-formed question, the time is right for an answer. Holding back info at the wrong time is just as treacherous a pitfall as dishing out too much. When a student asks an articulate question a couple of times, especially with an urgent note in her voice, it is not the time to play guessing games. That little blossom of a question is

plenty strong and deserves a good watering; otherwise it may die of drought.

I once worked with a teacher who believed "the discovery method" to be beneficial, but had a rather shallow understanding of it. With the admirable goal of letting students discover the concept at hand, he'd hold back virtually all information until his students were ready to throw him out the window. Worse yet, sometimes he'd buckle under the rising chorus of whining complaints and with great irritation say, "If you don't want to do the discovery method, we can just go back to straight lecture and textbook!!" His students were often so fed up with groping in the dark that they took him up on the offer and thus forfeited any chance of having a genuine learning experience with the universe.

So, offering information is not a bad thing. The problem comes when you and/or the student slip into the false state of reality where either of you believes that "the answer" can be conveyed in its entirety, to everyone's satisfaction, in a few seconds. As long as anyone holds this view, the interaction will be doomed to fail. Understanding does not come merely by the acquisition of more information. Any good answer will have several more questions wrapped up in it! A good answer will draw the student farther into the subject area and stoke the fires of fascination. Understanding will be gained slowly through repeated encounters with a set of concepts, with questions intertwining the whole process.

What we're looking for is a monumental realignment of our students' idea of learning. We want to model the idea that a question can linger in our midst comfortably, like an old friend, without stress, revisited again and again from various perspectives, each time letting go with a bit more insight and leading us to ever better understanding. Answers, in this new view, are merely preludes to more questions and more learning.

Sometimes a student or group of students has asked one or more questions several times, expressed them clearly, and is obviously keen to

---

2.   Jos Elstgeest wrote a very nice piece on good questions and answers called "The right question at the right time" in *Primary Science: Taking the Plunge*. Edited by Wynne Harlen. Oxford: Heinemann Educational, 1985.

take in more information. And sometimes you do have some good information or background to give about the topic. In this case, it may be time for a lecture. There is certainly nothing wrong with a lecture when both teacher and student have agreed to it. The source of information always needs to be evaluated, but if the teacher is honest and the student is skeptical, this is a fine way to learn. This can be part and parcel of a productive tinkering experience.

Here are my two steps of advice for giving answers, once you've ensured that the question is well rooted:

1. Give a single morsel of info, connected if at all possible to something the student is already familiar with.

   Student: "So, why does the pitch go up as the drum head tightens?"

   You: "Well, the same thing happened with the string on the guitar we made, right? The pitch of a sound is its frequency, that is, how fast the vibration is going back and forth. If something is tighter, it will tend to go back and forth faster, yes?"

2. Encourage more investigation, using stuff on hand.

   You: "Now see if it happens with this rubber band: stretch it loosely and then tighter, and see if you can see or hear the difference. You can also check to see if changing other factors would also raise the pitch."

Those two steps are often all you need for a while, maybe the whole class period. Soon you may be hearing additional questions, offsprings of the original. Deal with them each in this same manner.

With your answer, you want to avoid at all cost conveying the idea that since the student has received your morsel of wisdom, she now understands this concept and can rest at ease. After all, "It's always more complicated than that!" That is a famous quote from my friend and mentor, the director of the Exploratorium Teacher Institute, Paul Doherty. He doesn't say that just to cover his butt; it is critical that we all understand that whatever insight and information we gain right now here today is but a single stepping stone in the broad universe of questions. When he says it's more complicated than that, he's saying *don't get cocky, don't think you've mastered this:* your grasp of the phenomenon in question can always stand to be tighter, broader, or deeper.

On the other hand, when you hear a question, you may not know anything at all about the subject; in fact, some of the *students* may be ahead of you! Do not despair. The trick is to position yourself together with the students *in the midst of the tinkering*. We're all learning together, right? This may be a lame statement if you're the professor and students paid big bucks to be attending your class, but that is simply not the situation when you're doing tinkering. Heck, I've got a physics degree from MIT, and yet I routinely learn new things from farmworkers' kids about the common materials we employ and how they can function together to do amazing things. One of the golden aspects of teaching with tinkering is that you never know what the students will learn.[3] Given this, it is ridiculous to expect the teacher to be able to address each question that arises with a lucid explanation.

You as the facilitator may need to practice responding to questions when you are not sure of the answer. Go into the bathroom, alone, at night, and look at your eyes staring back at you from the mirror. Now repeat after me: "Wow, what a fabulous question! I don't know!"

---

3. This is one reason that, with disastrous consequences, tinkering has so rarely found its way into official sets of learning standards. You can almost see the educrats wringing their hands: "With so many learning possibilities available when tinkering, how will we ever determine what the students are supposed to learn? And what will we do, heaven forbid, if they learn something — gasp—that was not among our goals for them??!! No, much safer to stick with the bucket theory of education, that is, the student's brain is an empty bucket, and teachers should follow our step-by-step script in order to fill it with good, pure information." Sigh.

I must say that one of the biggest joys in my life is when a student asks me a question I don't have the answer for. I look forward to this, and cherish it each time it happens. It means good, good learning is on the wing.

When the question is well rooted but you don't know an answer and have stated as much, the next step is critical: *encourage further tinkering*. Accompany them if possible, at least to start out.

Student: "When I cover the open ends of these drums, the sound changes a lot. But when I cover this one, it sounds just the same? What's going on?" You: "Wow, brilliant observation. I'm not sure either, but let's look at the other factors. The two drums are not the same size, and are also not of the same material. Maybe we can make two of them exactly the same then change just one factor at time. Let's do it!"

Of course, you can always have your students pack up their questions at the end of a tinkering session and send them off to find answers elsewhere. But you likely have sources of information beyond what your students have. You can always get back to the students once you've had a chance to consult these sources—Paul Doherty always lets teachers know they've got a couple of days to get more info to the students. No one promised kids instant enlightenment! What's more, a well-rooted question will help focus ongoing learning. Every PhD candidate's got one! Professors and professional scientists and engineers and artists have many!

Thus, cherish your questions, question your answers, and take on learning for the long term; this is the way great tinkerers through the ages came up with all the good stuff. It's a kick, too.

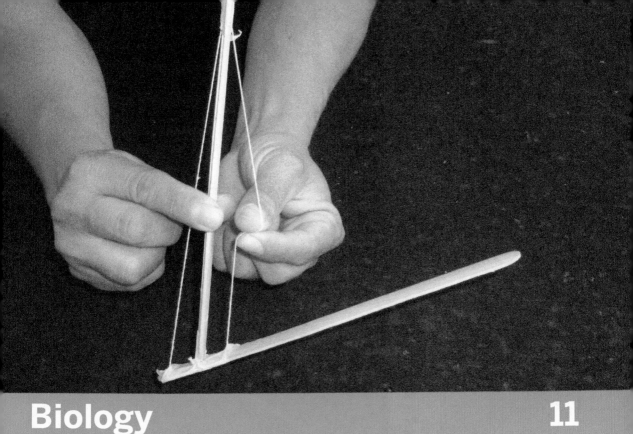

Biology 11

Exploring the Arm and Ankle Model

Tinkering with biology may bring to mind the unpleasantness of Dr. Frankenstein, cloning, or plastic surgery. But in the end, to learn the way things work, you do need to tinker with them, and this is certainly true with our own bodies. Throughout history, people have had to go great lengths to get permission or get around laws and tinker with dead people. To the extent that they were able to do this tinkering, real knowledge about body systems was slowly amassed.

Today you can easily leave your dead body to science, and in many cases it will be dissected by medical students. You won't find a surgeon who got her qualifications without completely tearing apart a corpse and observing each organ and tissue. If I ever go under the surgeon's knife, I'm going to hope I'm not the first hunk of meat the surgeon has ever chopped into.

Dissections are great ways to tinker with biology, and many great dissections are possible with nothing more than a trip to the supermarket or meat or fish shop, a razor blade, and a piece of cardboard for a dissection tray. Search the Internet or texts for a labeled photo of what you're dissecting and start slicing.

Another way to tinker with the way bodies work is to build models of body parts. If you get the model working similar to the real one, you'll gain insight into how the real one works. Our endoskeletons have muscles pulling on each bone. Tendons connect the muscles to the bones and ligaments link the bones to one another. Though all these parts are made of cells, the difference between these types of tissue is as large as the difference between string and bamboo. Making models of limbs brings out this point effortlessly. Dissections can be done together with building these models. After building them, taking apart a complete chicken wing or leg or foot will nicely demonstrate the way things work with real tissue.

Make: Arm Model

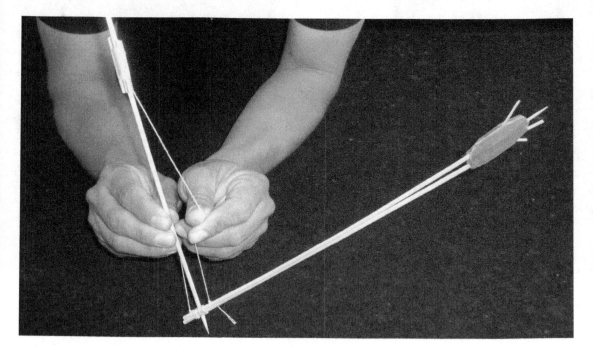

Figure 11-2. *The arm model in action*

Gather Stuff

- Three bamboo skewers
- Kite string
- Masking tape
- A bit of cardboard
- Toothpicks
- Straws
- Skinny rubber band

Gather Tools

- Scissors
- Side cutters

Tinker

Step 1

Grab two of the skewers together in parallel and wrap a rubber band around them, medium tightly, near the nonpointy end.

Step 2

Skewer the wrapped rubber band with the pointy end of the third skewer. Make it slide down on one side of the two skewers together (not between them).

Step 3

Tie one end of two lengths of kite string around the top one of the two parallel skewers. Slide the knot so that it's positioned near the rubber band.

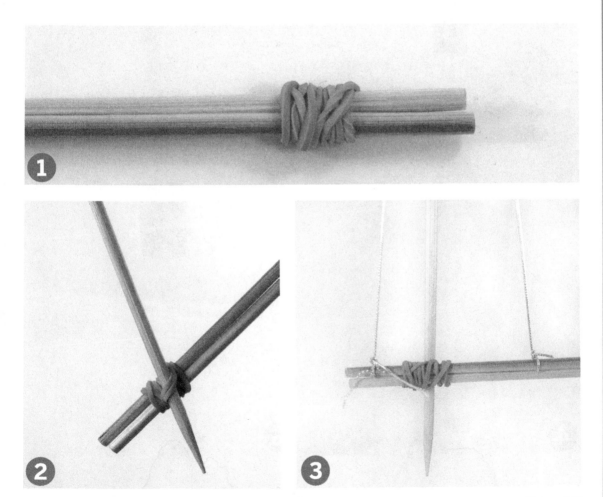

Step 4

Cut two short lengths of straw and tape them as shown near the top of the single skewer. Run the strings through them.

Step 5

Tape the two strings as tightly as possible to the top of the single skewer. You're going to be pulling on them, and you don't want them to come loose. This photo shows how we even hot glued them.

Step 6

Form a hand with the cardboard. The fingers could even be toothpicks. It's nice to slide the skewers into the corrugations of the cardboard and then leave them loose, not glued together, so as to allow arm-like motion.

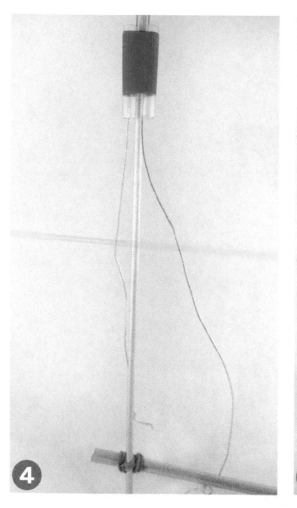

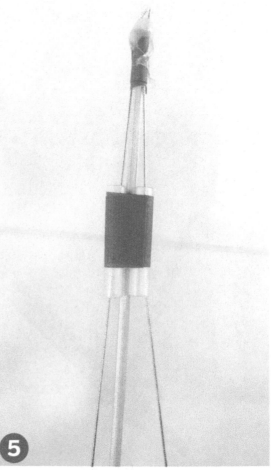

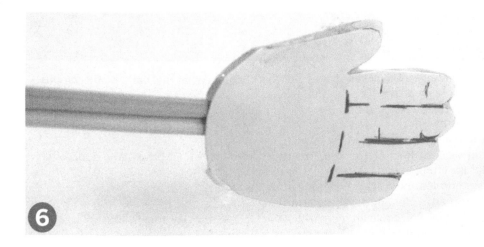

6

Step 7

There is your arm. When you tweak the front string, the arm raises, as in the first photo of this section (*Figure 11-2*).

Step 8

Tweak the back string, and the arm will move downward.

Step 9

Label the two strings if you want.

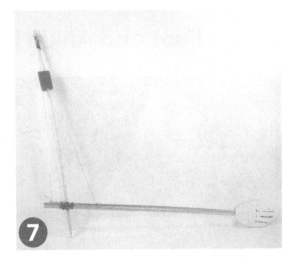

7

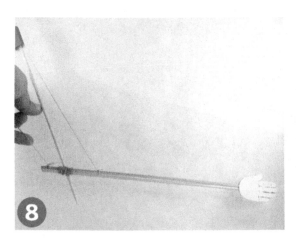

8

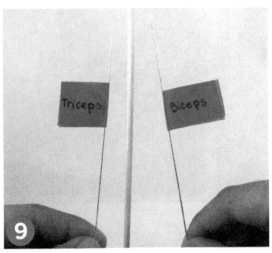

9

Check it out

- Look at the drawing of the arm at the end of this chapter. What real parts do the various parts of the model represent?
- When you pull on the front string, what kind of real life motion is it like? How about when you pull on the back string?
- How do you think the muscles do the pulling?
- What if you slide the knot on the forearm farther from the rubber band? Would this offer any advantage lifting things?
- What other muscles and parts are missing from this model?
- Would a longer arm offer any advantages in lifting?
- Which do you think you could lift more with, your biceps (front) or triceps (back)?
- Which do you usually lift with?

Make: Foot and Ankle Model

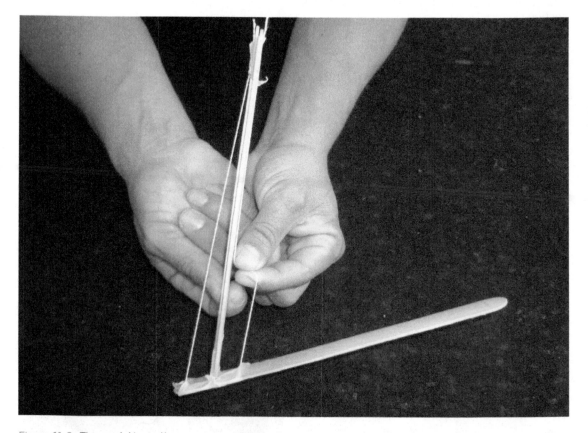

Figure 11-3. *The model in motion*

Gather More Stuff

- Three tongue depressors

Tinker

Step 1

Crack one of the tongue depressors the long way.

Step 2

Hot glue one of the cracked halves to the center of a whole one. These are the two bones in the shin

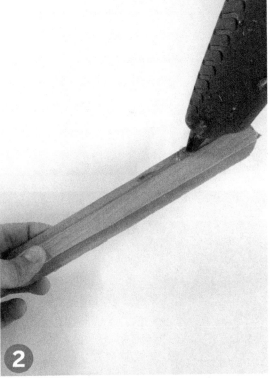

Step 3

Tape the end of the two glued pieces to the third one, near the end. When the glued pieces are standing upright, the cracked piece should face the short end of the third stick, as shown. Tape it front and back such that it makes a little hinge.

Step 4

Tape the tape hinge around the tongue depressor so that it is solid.

Step 5

Tape on a piece of string as shown. Put a knot in the end of the string so it doesn't easily slip out.

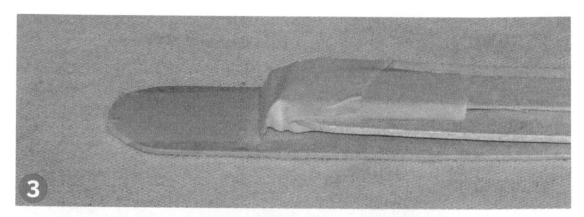

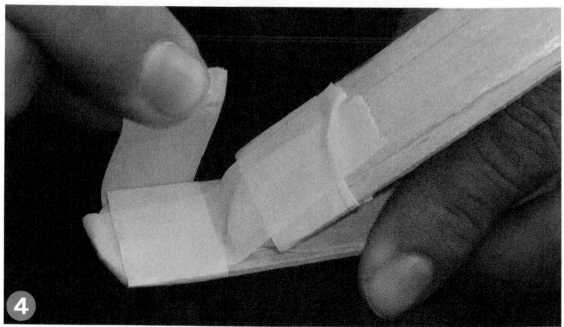

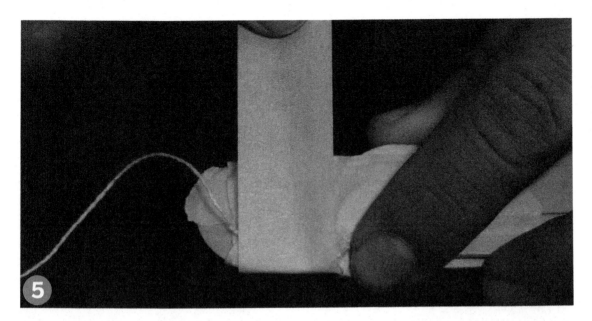

Step 6

Do the same in front as shown.

Step 7

There's your ankle. When you tweak the front string, the foot will rise as shown in the first photo of this section, *Figure 11-3*. When you tweak the back string, the foot will go down, pushing off the ground.

Step 8

Add labels if you want.

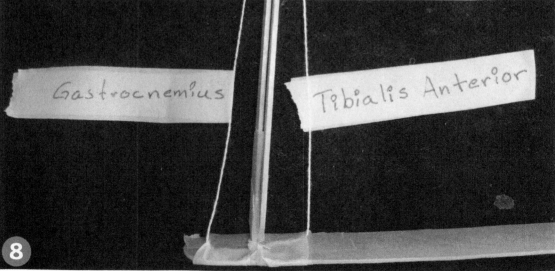

Gastrocnemius   Tibialis Anterior

- Look at the drawing of the ankle at the end of this chapter. What real parts do the various parts of the model represent?
- If you hold it just right, can you imitate a running motion?
- Which muscle is bigger on your leg? Can you work out why that is?
- What if the foot was longer or shorter?
- What if the part sticking out at the heel was longer?
- Can you figure out a way to make it more realistic? Skin and toes and everything?

What's Going On?

With a few exceptions (eye lids, tongues, hearts, and esophagi), our body moves by muscles pulling on tendons attached to bones bending at joints. The bones are levers, each with three parts: fulcrum (where it pivots), resistance arm (span between fulcrum and what you're lifting), and effort arm (span between fulcrum and where your muscle is pulling on it). How much force you give and how much force your muscle needs to provide is determined by the relative lengths of these arms. It's called *kinesiology*, baby!

Can you see where your muscles are hooked to your arm bones? Tense them up and follow them with your fingers. Try to work out which muscles pull which bones. In theory, you should be able to tense up each muscle separately, but since you don't usually use them that way, they may all tense up together.

Levers come in three types (*Figure 11-4*), one with the fulcrum in the center, and two with the fulcrum at the end. Most of our muscle-bone configurations are of the first and third types. It can be fun to explore the differences between these three types of levers, but it's not so critical. The key issue is that due to these lever arrangements, our muscles generally have to give more force than the weight of what we're lifting. The benefit is that we can move things farther than the distance the muscle actually moves, which is usually a very short distance. In physics, there are many trade-offs like this, and in the end you never get something for nothing. You'll never find a lever, for example, that multiplies both force and distance.

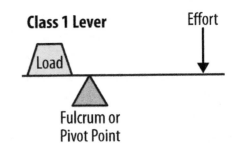

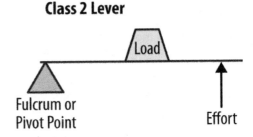

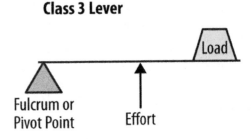

Figure 11-4. *Three classes of lever*

Did you find that when you tweak the string of the models a short distance, the result is a longer motion of the sticks? Can you see that this is the same thing happening in your arm and ankle?

Muscles are amazing. They're physio-chemical motion-makers in the biological world. When a muscle pulls (and they always pull, never push) millions of molecules, smaller than microscopic, are changing shape, cocking their heads in a very real manner, which results in the shortening of the entire tissue. This is what gives the force that can lift everything from a fork to a refrigerator: the changing shape of molecules and their bonds.

Watch one of your main arm muscles contract when you lift something. What do you figure is the total distance the muscle shrinks?

The strings in the arm model represent the *biceps* (front) and *triceps* (back) and the tendons that attach them to the bones. In the forearm, there are two bones, and the biceps muscle is attached to the one called the *radius*, more or less on top. The fulcrum of this lever is at the end and the effort in the middle, so this is a class 3 lever. According to physics, you should be able to lift twice as much when you lift a weight hanging from halfway down your forearm. Also, short arms will be able to lift more with the same muscle force. Of course, in both these situations, you won't be able to lift as far.

Try to lift a heavy bag first grasping the handle with your hand palm up in the "curl" motion. Then hang the bag halfway up your forearm and try lifting it again. Can you feel the difference?

The string in back is the triceps and pulls on the stubby elbow end of the other bone, the *ulna*, more or less on the bottom. Now the fulcrum is the middle and the effort is at the end, so it's a class 1 lever. We use our triceps to press down on the table or to push something forward. Most people are weaker with their triceps. Usually we lift things with our biceps, so they may get a more regular workout.

Put your arm up over your head and lift a heavy bag behind your back using your triceps. Body builders call this the French curl.

Many muscles of the body come in pairs, with one acting as the retractor and the other the power stroke. The biceps and triceps are an example, as are the *gastrocnemius* and *tibialis anterior* (see the diagram below). Sometimes the two are nowhere near the same size, since one does all the work. The gastrocnemius (the word comes from stomach, since your calf muscle looks like the belly of your leg) gives the power stroke for running, pulling up on the heel with a class 1 lever arrangement that pushes down hard on the ball of the foot.

Push yourself up on your tip-toes and let yourself back down. Do it 25 times. The pain you feel is your gastrocnemius!

Deer, cats, and dogs have a much larger distance between the fulcrum and the point of connection of the gastrocnemius. (The more I say that word, the better I like it.) This changes the mechanical advantage of their ankle lever, and puts considerably more spring in their step, so to speak. Just another reason you should not count on outrunning a jaguar or a hyena. These models can be made more realistic by adding a skin-like cover, perhaps a small sock or cloth tube from a scrunchy.

Arm:

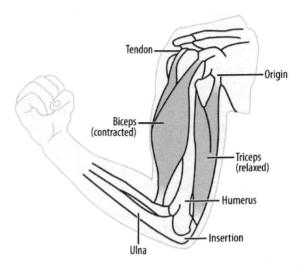

Ankle:

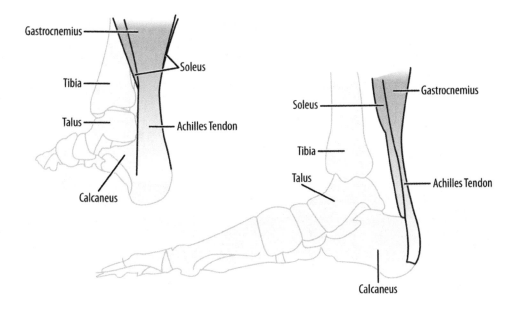

Dog's leg:

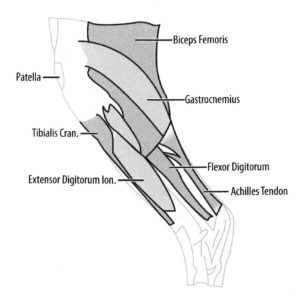

Cat's leg:

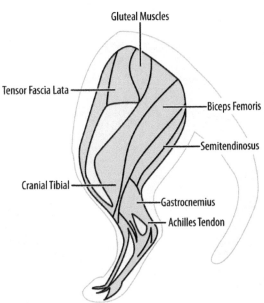

Keep On

We can make various other models related to the human body: Figure 11-5 shows a finger. The string is tendon, bits of tongue depressors are bones, tubes are carpel tunnels, and your pull is the muscle. This one is made with a drinking straw as the tubing.

The straw on the back springs it back into open position. In a real finger, there is another tendon back there (Figure 11-6).

Figure 11-5. *A finger model*

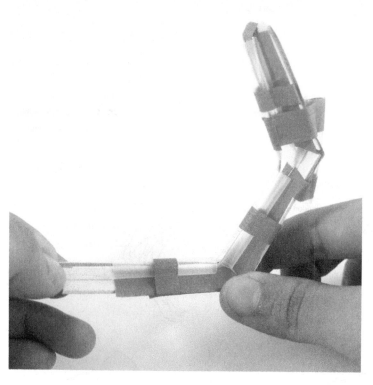

Figure 11-6. *Check out the tendon*

Figure 11-7 shows an eye model built around a mailing tube. The lens is adjustable—back and forth—so that you can focus on things at different distances (Figure 11-8). The retina in back is wax paper. The black paper is to keep other light from washing out the image on the wax paper.

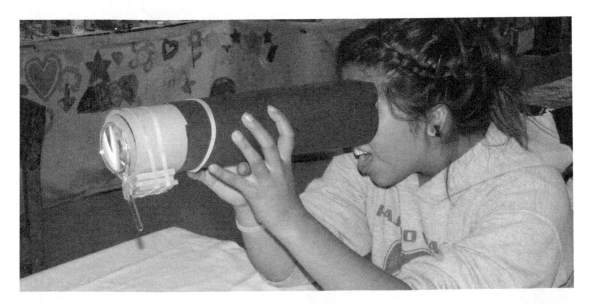

Figure 11-7. *The eye model*

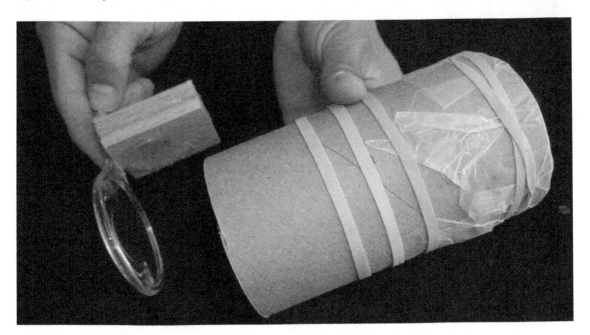

Figure 11-8. *The lense and the retina*

Finally, *Figure 11-9* shows an ultra-simple version of the classic balloon-in-the-bottle lung model. That's a half-balloon acting as a diaphragm muscle, taped around the chopped-off bottom of the bottle, and another balloon stuck in the neck as a lung. (Take the balloon out of the neck, and you've got an air gun.)

If you tinker with a cow or pig lung (*Figure 11-10*), you'll find thousands of little balloons that you can blow up with a straw.

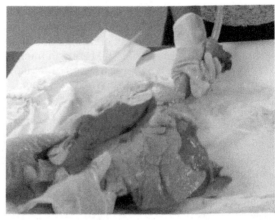

Figure 11-10. *A real lung!*

Figure 11-9. *Lung model*

Dan Sudran of the Mission Science Workshop in San Francisco has gone a bit nuts about bones over the last 10 years. He's collected and prepared hundreds of them, small as a mouse and large as a blue whale. He's got class sets of gopher skeleton puzzles. He's got a panel of femurs from mammals of all sizes. He's got entire ostrich and entire cow skeletons. Dan and I and several dozen of our nearest, dearest friends collected an entire baby gray whale skeleton off the beach near his house, and he got permission from the marine mammal authorities to put it on display for educational purposes. Dan's vision was this: if you put up a perfectly articulated whale skeleton, some people may come to be impressed by it. If on the other hand you pack it into tubs in the back of a delivery truck to be taken out and put together like a puzzle by groups of kids all around the area, much more real education will happen. He did the latter. Thousands of kids have taken part in the baby gray whale puzzle.

Vertabra puzzle:

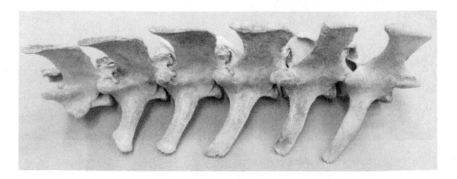

The weasel puzzle:

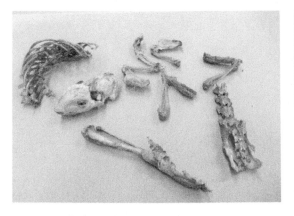

The articulated cow he found in a buddy's pasture:

Baby gray whale spine being put together:

Figure 11-11. *Whale panorama*

Internet Connections

- Search "tree art" or "tree sculpture" to see how you can (with patience) tinker with plants.
- Check out the Exploratorium's Mold Terrarium activity, and images of others, to see how you can tinker with fungus. *http://www.exploratorium.edu/complex ity/exhibit/mold.* html shows an exhibit that was once on the floor of the Exploratorium showing mold and other microbes spreading across a vast landscape under Plexiglas.
- You can watch many great videos under the topic of "squirrel obstacle course." This, like helping a hamster work its way through a maze, is tinkering with animal behavior.

Standards Topic Links

- Anatomy, human biology, body systems, organs, muscles, bones, and kinesiology

More Tinkering with Biology

- Karen Kalumuck, *Human Body Explorations* (Exploratorium Store, 2000; handson investigations on what makes us tick)
- Theo Jansen, *The Strandbeest Kit* (Gakken, 2011; Jansen is an artist-physicist who made a mechanical creature that moves like it's alive)
- Robert Bruce Thompson, Barbara Fritchman Thompson, *Illustrated Guide to Home Biology Experiments: All Lab, No Lecture* (Make:, 2012; brace yourself— this one's serious)

12 Standards and Assessment in the Tinkering Environment

Tinkering is not difficult to align to a set of standards, and assessment is also entirely possible. More and more, those creating standards and assessment have hands-on tinkering in mind.

Standards

I love my California State Science Content Standards; they encompass nearly every bit of knowledge learned since the dawn of creation. With six to eight years of hard study, any respectable PhD candidate should be able to master this information. It seems like a tall order, though, that we teachers have the solemn expectation upon us to impart it to each of our students, regardless of what they came in the door knowing.

The problem with content standards (aside from being drastically overstuffed) is that they are inevitably a list of bullet points, a set of info-snippets, each carefully crafted to encapsulate some critical morsel of our understanding of how the world works. In reality, to truly learn that information, one must have a personal, authentic experience with the concept in question. The experience will seat this new knowledge among the known knowledge already in the noggin and make the connections necessary to retain and contextualize it, as well as to get it all warmed up and ready for future use.

This is my view from experience, and it is also the view of the folks at the National Research Council, who put together the Framework for K–12 Science Education (2011), which forms the foundation for the New Generation Science Standards, a state-led initiative to make standards more functional and up to date. These new standards may be rolled out in some states by the end of 2013; check to see if they're coming your way.

Three dimensions compose the Framework and Standards, the first of which is "Scientific and Engineering Practices." "Practices," that is to say, don't just sit there and talk about it! This dimension is all about actions, such as:

- Asking questions (for science) and defining problems (for engineering)
- Developing and using models
- Planning and carrying out investigations
- Analyzing and interpreting data
- Engaging in argument from evidence

According to this framework, then, until the student is actively doing these things, she ain't learning well enough. I agree.

I've been doing science education in California for 20 years or so. At some point in the process of assembling the current list of state standards, California education leaders also realized the importance of real investigation and experimentation. I think it was the afternoon before they were due to be published, because what they did was to tack on a *separate section* (one could almost call it an addendum) at the *bottom* of each grade level's bullet list of content standards, which reads, identically, year after year:

Scientific progress is made by asking meaningful questions and conducting careful investigations. As a basis for understanding this concept and addressing the content in the other three strands [branches of content], students should develop their own questions and perform investigations.

Well. You couldn't do much better than that in laying out what real scientists, engineers, and technicians need in terms of good preparation. I'll translate it, in case your edu-speak is not up to the stratospheric level of the educrats here in California:

Let them tinker with real stuff! Then talk with them about their questions!

This glorious (though tacked-on) statement is followed by yet *another* list of bullet points, sometimes indecipherable, giving suggestions on what students should be able to do in this category, such as, "Interpret events by sequence and time from natural phenomena." These are mostly fine suggestions, and with a bit of translation, can be morphed into the fabulous tinkering activities this book is about.

The only dilemma is that the year's schedule is already bloated with the abovementioned science content factoids, *and* none of these experimentation and investigation bits are likely to turn up on a test, so most teachers ignore them, and students suffer. We are left to hope that this critical little section can be more vitally incorporated in the next round of standards, and as noted above, the prospects seem good.

The thing about content standards is that they're unavoidable. Like the cycle of the seasons, the motion of the planets—well actually, more like death, taxes, and toast landing jam side down, teachers these days have to try to "get through" their content standards. Heck, in California they make us post them *on the classroom wall*, presumably so little Angelica could pipe up and say, "Excuse me Maestro, the semester is slipping away, and we haven't yet covered standard 6.i., which reads: *Students know how levers confer mechanical advantage and how the application of this principle applies to the musculoskeletal system.* Shall we miss recess this week in order to master this?"

But really, it makes good sense to have kids throughout a region all on the same topic in the same year, if only so that when kids move schools, it's an easier transition. Plus, teachers do need guideposts for what to do on a week-by-week basis in their classroom. So here's what you do. Get to know your standards well, ugly and crude though they may seem, and especially know the basic categories. Then you find work-able tinkering activities that have a connection to those categories. After some satisfying sessions tinkering, students' questions will be hovering in your classroom like flies after a dissection. These questions will undoubtedly have links to your set of standards. Now's your chance: as you discuss and nourish and valorize these questions, you can insert, discreetly, the factoids from the standards. You're golden.

(Inevitably there will be misfit concepts from the standards that have little connection to anything in the realm of normal human experience. Those will be special precisely because they are isolated. Print them out on little cards and tell the students that they may not see any connection here, but that sometime in the future this info may come in handy, so learn it now.)

As an example, take the first activity in this book. Tinkering with sound normally fits into a category of physical science concepts having to do with waves and vibrations. In the current California State Science Standards, we see:

- A single line about sound in second grade (under the heading of motion—a fairly abstract representation of sound for that age level)
- A brief mention of sound among other waves in third grade
- An almost incomprehensible grouping of sound in the category of thermal energy in sixth grade
- A nice bit about the structure of the ear in seventh grade
- A single mention about sound as a wave in the collective high school physical science standards.[1]

Oddly, the eighth grade standards do not mention sound or waves, though students are meant to be entirely focused on physical science that year. But certainly one can easily relate sound to other points on the eighth grade list: motion, velocity, energy, force, and so on. Thus you could do sound tinkering in second, third,

---

1. It is troubling to find no direct mention of the ear in high school standards, an age when nearly every student is channeling many hours of music each day directly to their inner ears, much of it at dangerously high volume levels. Regardless of public health considerations, it is clear that sound and music are deeply significant to kids and thus are obvious topics for science class or tinkering.

sixth, seventh, or eighth grades, and anytime in high school, and be "teaching to the standards." Furthermore, I have had great success with the arguments that (1) my students clearly missed that particular standard when they should have got it last year and (2) I happen to know that the teachers next year will not be addressing that standard, so I'd better do it this year. Now grades 1 through 12 are covered.

As I mentioned, you should glean the key language snippets from the official standards and offer them to the students as you chat with them about their tinkering. Here that means, "... sound is made by vibrating objects and can be described by its pitch and volume" and "...energy can be carried from one place to another by waves," and so on. All these formally condensed points are easy to bring to life through the activities and the questions that always arise from them. In stark contrast, to deliver this sort of concise and sterile language to students without the chance to experience the concept is a fool's errand.

After-school programs often don't have the pressure to teach to the standards, but some do, and they are not often science standards at all. Many schools in our area can't get students up to language and math benchmarks during the school day, so after-school programs are feeling the pressure. Many have adopted remedial catchup materials that try to mechanically hoist Juan up to where his peers are. To this I say, "Resist!" Pedantic, soulless, rote learning during the school day didn't give Juan the language skills he needs, so why should he be inflicted with more of this agony after school?

You can easily flex your tinkering plans to address this critical and spreading issue. What Juan needs is an authentic experience fooling around with real stuff. Then chances are high he'll be primed to learn some language skills to communicate about it. He can write about it, talk about it, or do activities exploring the language involved in it. This takes some effort but is well worth it.[2]

I'll close this topic with a powerful story of one my main mentors, Phillip Morrison. An award-winning MIT professor, longtime book editor for *Scientific American*, and creator of various acclaimed TV programs promoting public understanding of science, he spent many summers of the 1980s and 1990s at the Exploratorium as a visiting educator. One summer I was there when he and his wife-partner Phyllis were presenting a series of teacher workshops on hands-on astronomy, and I attended a brown-bag lunch presentation of his.

The topic was a bit of a shock: the plateauing of he earth's population growth curve, which had been recently predicted by mathematical analysis. Perplexed, I later approached him. Being a friend and mentee, I was direct: fascinating subject, but how in the world did you choose it? Astronomy was on the menu, no?

His immediate reply, given with his trademark smile of confidence, has guided me for decades since then. He said, essentially, that he chose the topic because it interested him, and because it did, he knew he could interest others with it. There are infinite, valuable topics to learn about, and everyone wants to learn from people who are fascinated by what they're teaching. In fact—he didn't hesitate to add—if a teacher winds up in the ugly position of having to teach something she's not interested in, it's a bust: meaningful teaching and learning will not happen. And if you are that teacher, there is no ethical remedy but to quit.

In other words, teachers and learning facilitators must have the leeway to design and shape what they offer in ways that excite them. In the end, freedom for the teacher serves the student. So use your standards; don't let them use you.

---

2. The Exploratorium Institute for Inquiry has been working on a project that proposes to do just this with schools in Sonoma Valley, California. It is funded by the Department of Education and is meant to last through 2015. Another such effort here in the Monterey Bay area in the late '90s was called Language Acquisition in Science Education for Rural Schools (LASERS). Finally, the Australian Academy of Sciences has produced an extensive set of excellent teaching materials for primary schools called Primary Connections that uses experiential learning to link the sciences with literacy.

Assessment

The role of assessment within the educational process is of utmost import...

Wait a minute. Why are we so concerned about assessment? What do we want to do, put our student's brain through an MRI to determine exactly which fragments of our ultra-critical knowledge have been absorbed into her gray matter? Or do we fantasize about some sort of sci-fi tool that we can poke on her forehead and use to assess precisely whether she retained the last 13 info-bits we spewed at her?

Look, let's give it a rest. She has been assessed on everything from adjectives to the associative property ever since she graduated from diapers. Perhaps she'd be more motivated to learn something, especially something that interests her, if we'd lay off with this !@#$%^&* assessment neurosis. Already she's probably developed an embarrassing knee-jerk reflex to regurgitate onto a quiz paper immediately after each baby step that happens in her formal education.

So I say we forget about our beloved assessment. Use the time to do more tinkering. Students will love us for it, and we'll live longer as well.

Hold on, I'm sorry. It *is* important to figure out if our students are learning anything or not, primarily as an avenue to help them do the best they can. But I firmly believe that today's assessment obsession is nothing short of damaging to students' development, not to mention poisonous to confidence and self-esteem.

I've seen more evidence than I care to relate of young children—kindergarten and first grade—already convinced of their lack of intelligence and potential to learn something, thanks entirely to school assessment regimes. Here the school is not only *not* helping students by its assessment processes, but *actively* harming them. I think any assessment method needs to ensure first and foremost that it will not inflict this sort of harm.

By demanding some numerical readout on the minute-by-minute learning that is obviously happening in a tinkering session, we may do ir-

reparable damage to this picture of educational excellence. We may forever crush the easy, natural, primeval tendency of those students to dig into something they want to know about and learn more. At the very least we stand a good chance of stultifying genuine learning.

Tinkering is most often joyous, and any attempt at assessment should not snuff out this joy. I work primarily in the after-school domain, so I don't feel as much pressure in this area as I would if I did more school-day work. I won't permit normal exams or quizzes in the course of tinkering, but I also don't believe I have many especially brilliant ideas on how to assess tinkering. Here I'll lay out what we do, which has been effective for us and is all pretty normal stuff, all keeping the fires of joy alive:

- Accompanying students during the tinkering and talking with them to see how much they grasp conceptually.
- Plying them with focus questions ahead of time, then during the tinkering nailing down some answers to the questions, and in the end verbally quizzing them.
- Having them keep a journal and refer back to it from time to time.
- Having them complete an "evaluation questionnaire" when they've finished. See Appendix B for common prompts we use for their journals and questionnaires.
- Having them draw or photograph or shoot video of the project.
- Catching them on video explaining what they think they got from the project.
- Having them prepare and carry out presentations to wider audiences, such as the younger students in the next room, and monitoring the presentations to see if they're accurate.

Some of these things do not require a lot of extra effort and give a good indication of any content learned. We have made it second nature in our programs to carry out one or more of these methods.

The bottom line is this: we believe our students have much in their midst to learn when they're involved in good tinkering. The important thing is to give the awareness and responsibility of learning to *them*, and then encourage *them* to keep digging for more insight and understanding through their whole life. Our goal is not information transfer—this is cheap and readily available for everyone in the modern world—but rather to create and sustain rich, supportive educational environments that put students in striking distance of the skills, knowledge, and wisdom they yearn. Can we really do any better than that?

Engineering and Motors   13

Fun with a Hovercraft and Spin Art

You can tinker with circuits, magnets, and mechanics separately and come up with hundreds of great projects, or you can put them together and invent go-machines of all shapes and sizes. The trick is to get a hold of a little hobby motor.

A motor is an amazing thing. The first one was developed by Michael Faraday in 1821, just one year after Hans Christian Orsted accidentally stumbled upon the link between electricity and magnetism. Up until then, electricity, magnets, and motion were all separate areas of research and tinkering. Orsted noticed that electric current makes a magnetic field, and Faraday used that field to produce continuous motion. Don't take my word for it, but inside the motor, you'll find coils of wire forming electromagnets, and usually permanent magnets, too. The magnets in there push and pull on each other to turn the shaft.

And once you've got a shaft turning, baby, the world is yours! Look at the motors in our lives: washing machines and dryers, refrigerator compressors, CD/DVD players and computer disk drives, automatic cameras, automatic car windows, blenders, hair dryers, vacuums, fans, heaters and air conditioners, car starters, grain moving motors, heavy lift motors, conveyor belts, escalators and elevators, automatic doors and gates, drills, grinders, lathes, saws, milling machines, clocks, and, last but not least, toys.

Any toy with a battery that does something be-
sides sing and light up is likely to have a motor
inside of it. Sometimes it will be hooked into a set
of gears or cams, and other times its shaft will be
connected directly to a fan or wheel.

Cameras often have only a small battery in them,
but also may have circuits that boost the low battery voltage to a high voltage necessary to set off
the flash bulb. Be careful opening cameras! I've
never seen a toy with such a circuit.

Most of these motors run on 3 to 12 volts, a perfectly safe voltage to fiddle around with. So, rip
apart a dead toy and extract the motor, or go out
and buy one for a couple of bucks. Now the action can really begin. You can build cars, boats,
submarines, planes, and many things that just sit
and spin or vibrate in a fascinating manner. Here
are a few ideas, but really, the sky is the limit. I've
personally written an entire book on motor engineering projects and have developed enough
projects since then to write a second book.

Make: Hovercraft

Figure 13-2. *The hovercraft*

Gather Stuff

- Paper plate
- Flexible plastic cup
- Hobby motor
- Two connection wires, three to four feet long
- Two to four D batteries
- Masking tape
- Electrical tape

Gather Tools

- Box cutter
- Hot glue gun with glue
- Scissors
- Nail
- Wire strippers
- Side cutters
- Pen

Tinker

Step 1

Cut the cup in half.

Step 2

Set it on the upturned paper plate with the cup's bottom up against the outer circle on the base of the plate. Mark around it.

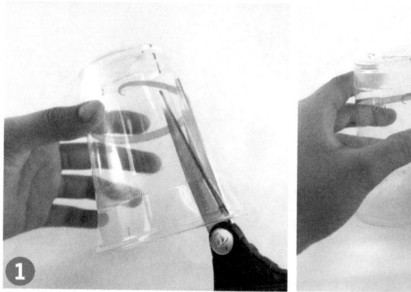

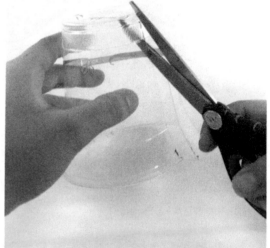

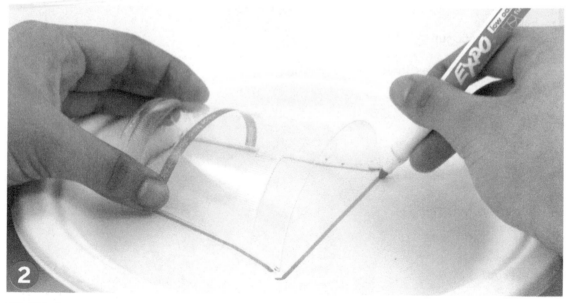

Step 3

Cut it out on three sides, leaving the one line corresponding to the bottom of the cup. Use the box cutter and scissors. Stick one piece of tape under the end of the flap in preparation for fastening it down.

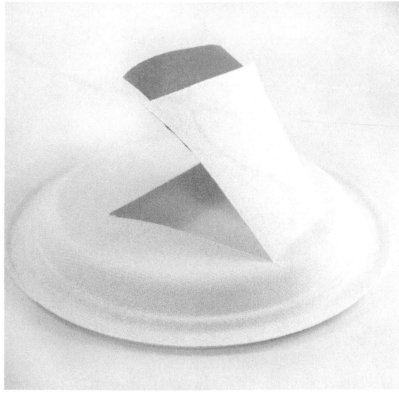

Step 4

Tape the half cup down to the plate. Tape the flap up inside the cup. This will direct the air from the propeller down under the paper plate.

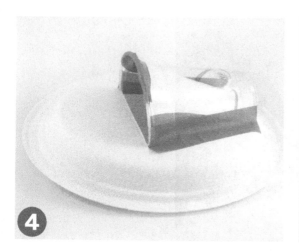

Step 5

Strip the two connection wires on both ends and connect them to the motor's terminals.

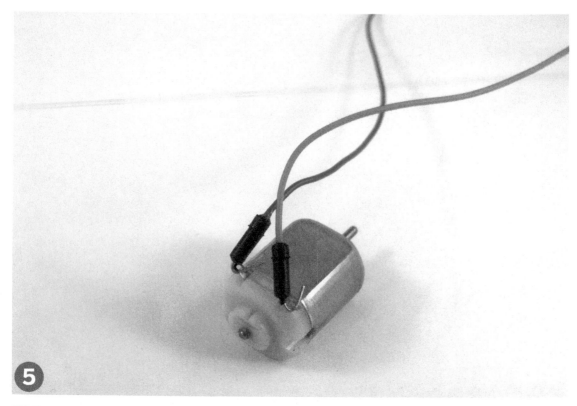

Step 6

Wad a bit of aluminum foil around the tip of one of the motor wires to make better contact. The other end has to be disconnectable to switch the motor off and on. Here you can see our unbent paper clip system for the switch side. Make a battery pack with two to four batteries and electrical tape. Press and tape the batteries and wires together securely.

Step 7

From the unused side of the cup, cut a propeller about two inches long, half an inch wide. Cut it at an angle on the side of the cup and it will have the proper curvature built in. Round the edges for smooth action.

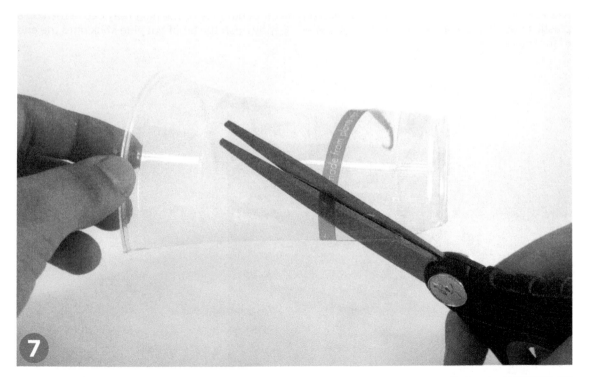

Step 8

Cut a bit of hot glue, maybe one inch long. Poke a hole in it with a small nail. It is important to poke the hole straight in: this is where the motor shaft will go in. If you'd like to make it easier, heat up the nail by sliding it into a hot glue gun's tip for several seconds before you push it into the stick.

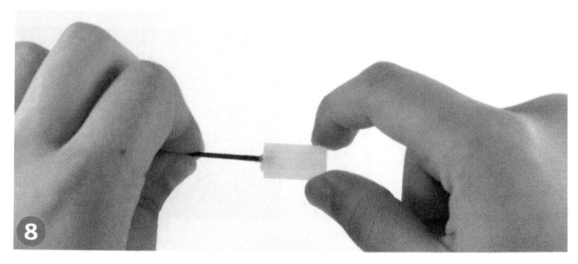

Step 9

Hot glue the propeller onto the nonholed end of the bit of hot-glue, smack dab in the center. It's best if the curvature comes down around the glue stick as shown, instead of curving up away from it. Wait a few seconds after you put the hot glue blob on the bit of hot glue or the heat may cause the cup's plastic to bend (but you can always tweak it later). Gently push the bit of hot glue stick onto the end of the motor shaft.

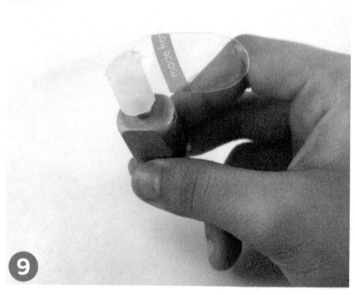

Step 10

Glue the motor snugly beneath the top of the cup. The closer you can get it, the less air will be lost around the side. Trim the propeller if it touches the paper plate when it turns.

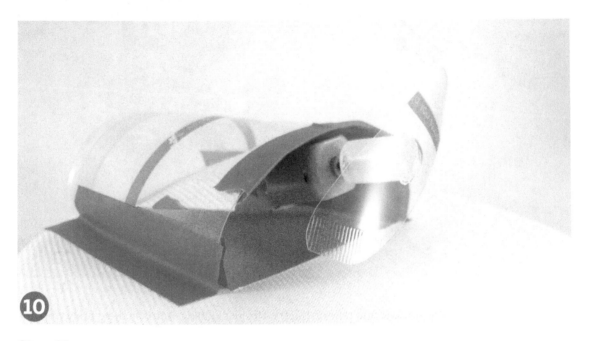

Step 11

Tape the wires back on the top of the cup to keep them out of the way of the spinning propeller.

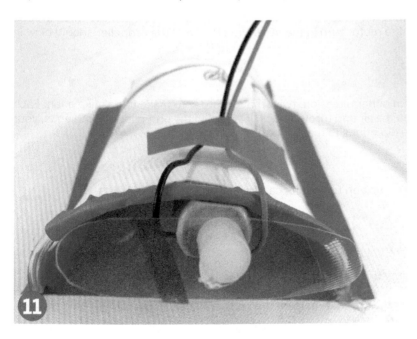

Let it rip!
It should go tearing around the table when you hook up the batteries.

If it doesn't, chuck it in the garbage and take up poetry. No, just kidding. Actually they rarely work the first time. Here are a few of the common things that go wrong:

The motor is going the wrong way

If you point it at your face, you shouldn't feel much air, whereas if you put your hand under the hovercraft, you should feel it blowing like crazy. If it's blowing the wrong way, switch the wires around on the battery. This will make the motor turn in the other direction, and the propeller should blow in the right direction.

The propeller is not bent properly

If it is not blowing much at all in either direction, twist that propeller to make it look like a fan. Each side should be twisted to scoop the air and throw it back. If you grab the propeller by both ends, your hands should twist in opposite directions to make it right.

Not enough batteries

More is usually better, though getting up to 12 volts can fry one of these little motors.

The table is not smooth enough

Find a smooth one, clean and dry.

The plate got bent in the building process

It needs to be pretty flat around the rim that rests on the table. You can smooth it with your fingers.

The motor should be hanging right above the center of the plate, too. If it's too close to one edge, rip off the motor and reglue it.

Your karma is no good

Do some good deeds and try again.

Check it out

- Is this thing really flying? Check out the distance between the edge of the plate and the table.
- What if the propeller was longer? Shorter?
- Fly it off the edge of the table. What happens?
- Can you make a way to control the direction?
- How is it getting motion both up and forward?
- Why don't you see these things flying down the street?

Make: Art Spinner

Figure 13-3. *The spinner in action*

Gather More Stuff

- Base board—any moderately heavy chunk of wood
- Another motor and battery
- Two more connection wires
- Small bit of wood
- File folders
- Paper plates
- Sticks of wood
- Liquid paint or food coloring

Tinker

Step 1

Strip all four ends of two long wires and connect them to the motor terminals. Glue the motor to the side of the base board with the shaft sticking up above the surface.

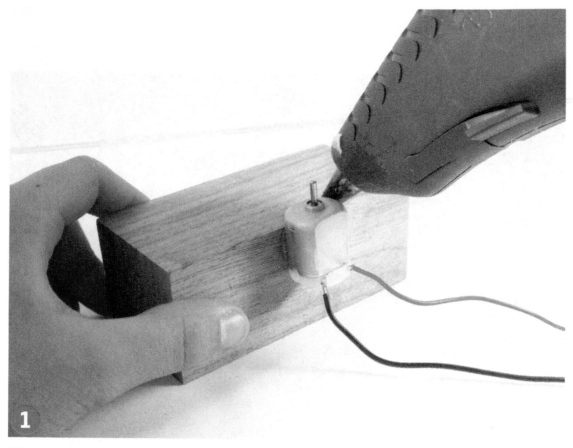

Step 2

Find a drill bit or nail to use in making a hole for the motor shaft. The hole has to be just smaller than the shaft so that the wood grips the shaft. It also has to be square with the piece of wood—straight in and straight out. If you can't find a drill bit, you can just tap a nail into the wood. You can just tap the nail into the wood and then yank it out, but we find it easier to put that nail into a drill and make the hole. If you find a nail you think will work, but then it turns out to be too small, you can smash it a bit with a hammer to make it flat, which will enlarge the size of the hole it makes.

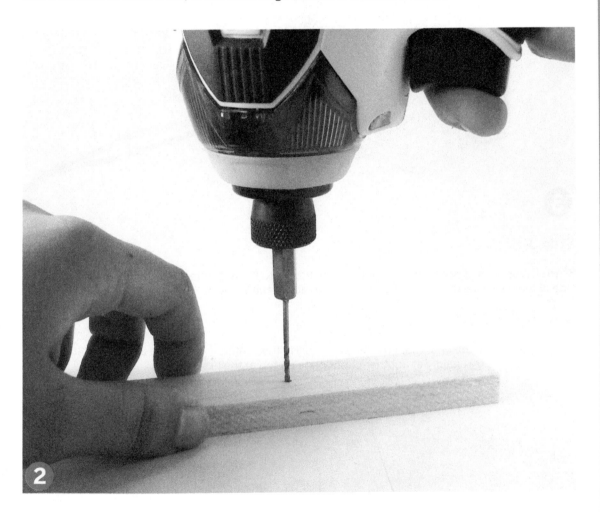

3

Step 3

Strip all four ends of two long wires and connect them to the motor terminals. Glue the motor to the side of the base board with the shaft sticking up above the surface.

Step 4

Hot glue a paper plate onto that little block *exactly* in the center. If you spin it and it's not in the center, rip it off and try again. Set another paper plate on top of the first one.

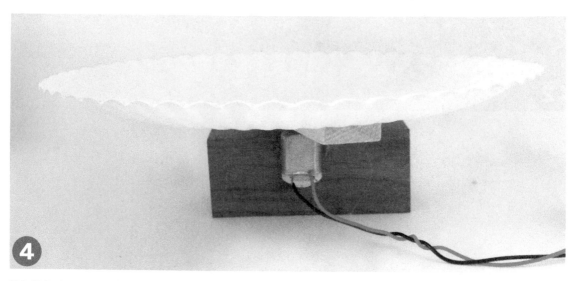

4

Step 5

Unless you want to paint your table, floor, walls, and self, build a little wall and floor around the spinner with file folders or thin cardboard. Now you're ready to spin some colors.

Step 6

Hook up the motor to the battery; you can just press the two bare wires to the two ends of the battery. When it gets going, drip the colors onto the plate little by little. When you're happy with one, slide the plate off, put it up on the refrigerator, and drop a clean one onto the base plate for your next masterpiece.

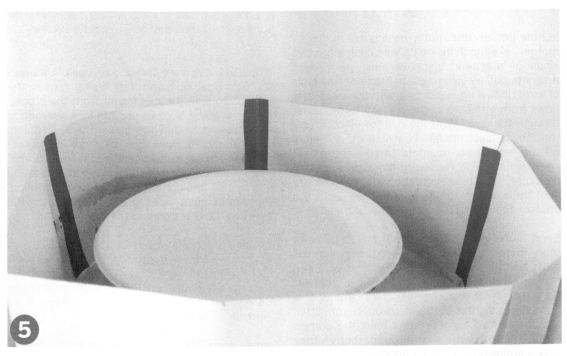

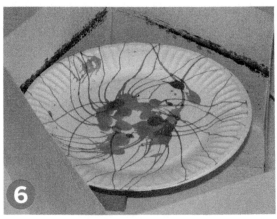

Check it out

- If you hook up more batteries and spin it faster, how do the patterns change?
- Why can you use a single battery on this one, but the hovercraft needs two or more?
- If you use larger or smaller plates, how do the patterns differ?

What's Going On?

Let me go over the motor basics again. If your motor is like most, inside its wee casing are two kinds of magnets: black ceramic permanent magnets and electromagnets made of tiny coils of wire. Hook it up in a circuit, and the electricity flows from one pole of the battery to the other through the coils. With the rush of electric current, these electromagnets switch on and push off the permanent magnets to spin the motor shaft around. The electricity flows through tiny contacts called *brushes* that connect and reconnect the electromagnets so that they're always pushing the shaft in the same direction.

Send more electricity through the motor with more batteries, and the electromagnets will push harder resulting in a faster spin. Send through too much electricity, and the poor little brushes will vaporize like the filament of a burned-out lightbulb. Send through too little, and the electromagnets won't have the oomph required to push the shaft around, though they may still slowly drain your battery.

Grab the shaft of a spinning motor. Can you believe that the force you feel is just tiny magnets pushing on each other?

Hovercraft don't fly; they float on a cushion of air. This gives them the significant advantage of nearly eliminating friction with the ground. Ground friction is what holds you back, so these gadgets can really go! Stopping, on the other hand, can be problematic, as can turning tight corners. That's one main reason you don't see many hovercraft waiting at the traffic light. But they work great on the surface of water.

Can you see a tiny space between the plate and the table? When my hovercraft flies over the edge of the table, that air cushion is broken and it skids to a halt.

Propellers throw air forward or back. This particular model of hovercraft cleverly uses the half cup to split the moving air from the propeller, some going back to provide forward motion and some going down to power the air cushion. Real hovercraft have one or more propellers dedicated to each of those two functions. Sometimes the propulsion propellers are able to swing side to side and control the direction of motion. Sometimes giant vanes are used to control the direction. Ours is doomed to random flight.

Can you think of a way to add another motor or control system of some kind that could guide your hovercraft? We tried using an electromagnet to push a vane back and forth, adding two more wires to the tether. It sorta worked, if we used our imagination.

Large propellers will move slower because they're pushing on more air. Since these motors are made to spin fast, you don't want too big a propeller. If yours is too small, on the other hand, it won't push enough air to generate enough force.

The spin art gizmo pushes the motor's lower speed limits. It has to start that (relatively) huge plate spinning and then keep it spinning when the paint splats onto it.

Did you notice how slowly it starts spinning, and how any little contact will prevent it from starting? During this start-up phase, the motor draws a huge current from the battery. That's one of the interesting quirks of electrical motors: if you physically stop the shaft, the motor draws much more current from the battery than when it's running.

Once the motor gets some momentum it does the job fine, and on only one battery. This is the way inertia works: it's usually harder to get something going than to keep it going. The same goes for the drops of paint when they hit the plate. After falling tranquilly through the air, they suddenly encounter the spinning plate. The moving surface of the plate gives the drop an instant push in one direction, and it begins moving in that direction. But the plate is spinning, so the next instant the force is in a slightly different direction.

Well, thanks to inertia again, the drop is not interested in changing directions, so it begins a curved path, partially succumbing to the torque of the plate, partially following its own inertial will out to the edge of the plate. The result is magnificent art.

Did your spinner give paint line forms like the ones pictured earlier? The spin art shown here was done with food coloring. If you use thicker paint, it may show less curving and more straight-out (radial) motion. To better understand this rotational motion, try tossing on beads, coins, flour, bubbles, or anything else you can think of. [1]

When a drop finally reaches the edge of the plate, it gets launched into the air. No longer feeling the torque of the plate, it will follow an arrow-straight path until it hits the wall, or the unlucky face of an innocent bystander.

Keep On

Many more tinkering projects are possible with this little spin master. One of the simplest is a car that is blown along by a propeller. The propeller is the same as in the hovercraft project, perhaps a bit longer. The car has to be pretty light, and the bearings have to be pretty good—any little friction will stop it from going. The bearings here are bamboo skewers—the straightest ones in

the pack—inside of drinking straws. Wheels are bottle caps, but could be poker chips or anything round. We use a 9 volt battery for maximum torque, but it won't live long. Two or three AA batteries would also do the trick and last longer.

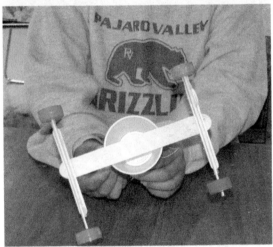

The unlikely looking contraption in *Figure 13-4* is a turning display table mounted on a hunk of two-by-four. It is one of our more difficult projects. These motors always spin quickly; there's no slow setting. If you give them only a bit of juice, they lose all force and stop. One day a kid wanted to make a slowly turning display table for the snazzy little model car he had just finished. I told him to forget it, given that these motors

1.   A large, slowly spinning turntable is a popular exhibit at the Exploratorium and other museums. You roll balls, disks and wheels onto it and watch their crazy trajectories. You can even put a small child on and see what happens; mine thought it was great fun.

only turn fast. He would not take no for an answer and kept nagging me until we came up with this. It's driven by means of the tilted shaft of the motor directly contacting a CD, which is held in the horizontal position by a dowel loosely stuck in a hole (*Figure 13-5*). The weight of the wood and other stuff on the CD (the model car) presses the CD against the motor shaft, and the whole thing turns. Since the shaft diameter is tiny, even though it's spinning like crazy, the CD goes round in lazy circles. Furthermore—and this is from my inner teacher—I rigged the motor with a binder-clip slider (*Figure 13-6*) such that you can slide it closer in to the center of the CD or further out toward the edge. This is a cheap transmission: the closer in it gets, the faster the CD spins; the farther out, the slower it spins.

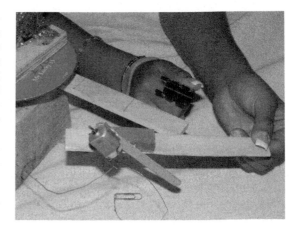

Figure 13-6. *The binder clip slider*

We sometimes fiddle around with proto-motors. These act sort of like a motor to show how motors work. *Figure 13-7* shows one built on an upturned cup, balanced on a pencil point stuck in a bead hot glued to the bottom of the inside of the cup. Four magnets are hot glued around the edge of the cup (*Figure 13-8*), each with the same pole out. We then bring in an electromagnet, which you can turn off and on. By switching it off and on at the right moments, we can make this little device spin around. Useless, but instructive.

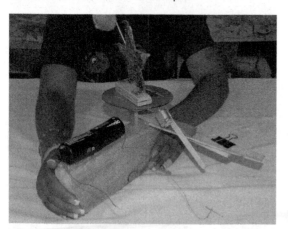

Figure 13-4. *The turing display table*

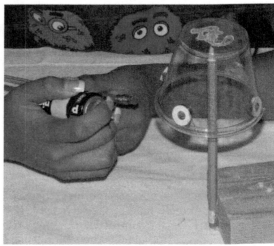

Figure 13-7. *A proto motor*

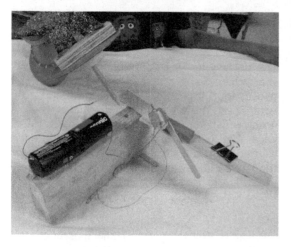

Figure 13-5. *A closer look at the device*

Figure 13-8. *Magnets and beads glued in place*

Of course, if you want to learn more about how these motors work, rip one apart (*Figure 13-9*)!

Figure 13-9. *Dissecting a motor*

Internet Connections

- Search for the biggest and smallest motors in the world, and prepare to change your vision of what a motor is and does.
- Check out a video of Faraday's motor, and other *homopolar* motors.
- Some extremely efficient vehicles have been developed with one or more wheels having a motor built right inside. Search "wheel hub motor."

Standards Topic Links

- Electricity and magnetism, motors, electrical coils, force, energy, power, torque, and magnetic fields

More Tinkering with Engineering

- I put a couple dozen of these motor activities, including a self-contained hovercraft that you can launch down a long hallway, into my first two books:
  - *Kinetic Contraptions* (Chicago Review Press, 2010)
  - *Stomp Rockets, Catapults, and Kaleidoscopes* (Chicago Review Press, 2008)
- Each of the commercial construction toys (Tinker Toys, Lego, K'Nex, Erector) includes a motor in the top-tier fancy set. Capsella toys were designed entirely around hooking up little motors inside capsules in creative ways; the system seems to exist now in the Danish company IQ-Key, but you can also get used ones on the Internet.
- The Maker Shed has several slick kits for motor toys. One good one is the SpinBot. Find them at *www.makershed.com.*

14 Final Notes

With the activities and pointers described here, we've scarcely scratched the surface. There's tinkering to be done in nearly every area of life, and a world of tinkering education to be had outside the official curriculum. The road lies open before us.

By this point, the more academically inclined of my readers will have grown quite frustrated by my unwillingness to offer a simple definition to what tinkering is. I've been beating around the philosophical bush since page one. What, after all, is the difference between tinkering and inquiry? Between tinkering and "hands-on science?" Between tinkering and the scientific method, or gathering empirical data? Can you tinker with chess or number puzzles? How systematically can you tinker before you're no longer tinkering?

While I'd love to discuss these questions with you someday on a pleasant stroll through the hillocks of the Shire, I can't think them too important. Tinkering for me is kids learning by making stuff, whether it be a model airplane, an electronic gizmo, a wooden chair, a leaf album, a mess of fish guts during a dissection, a rock candy crystal, a paper cup flower garden, or a discovery into the workings of their busted hair dryer. When I can facilitate kids tinkering with these things, I'm super confident that I'm helping them to approach and understand their world in positive, practical, constructive, and ultimately effective manner, as well as helping them get a firm, guiding grip on their own education. For some of them, this is nothing short of revolutionary on a personal level.

Hold fast to idealism! Yes, to become real engineers, your kids will doubtless have to trudge through countless, mind-numbing problem sets, but Michael Faraday received little formal education and rarely wrote an equation in the process of discovering the relation between light and magnets and making the first motor! He was a tinkering superstar! Sure, to make the next big bio-medical breakthrough, your students will likely have to mix hard-to-get genetic scraps with various highly purified enzymes thousands of times in a smelly laboratory with precisely the same procedure, but Anton van Leeuwenhoek was an amateur tinkerer who developed his own microscopes to become the best of the day and with them viewed for the first time single-celled organisms, muscle fibers, and capillary blood flow! The man knew how to tinker!

Think of the famous tinkerers you know and what they've given the world! Leonardo de Vinci, Robert Fulton, Marie Curie, Thomas Crapper (yes, you've guessed correctly), Frank Lloyd Wright and his distant kin Wilbur and Orville, Edison, Ford, Salk and Sabin, Cousteau, Hewlett and Packard, and Gates (tinkering made him realize he didn't need Harvard).

Zounds. They're all men but one, some of them quite sexist! But don't you believe for an instant that there were not a lot of women tinkering with the same stuff as those men every chance they got. The reason you don't know more famous women tinkerers has more to do with the miserably inferior opportunities given to women over the years than with any male predilection to tinkering. That, and the fact that men also held the power of the press, so all too often even when a woman healer or midwife made a big break-

---

1. The Exploratorium and Stanford Research Insitute have been doing research on these questions and others related to the broader educational movement around making and tinkering. Watch *http://www.exploratorium.edu/cils* for information regarding this research.

through, men would downplay or ignore it or even take credit for it. Certainly, today it is no problem to find women in every area of professional and amateur tinkering.

Perhaps even more significant than those famous names are the millions of tinkerers like myself who, despite thousands of hours tinkering like crazy, have added little to the sum total of human accomplishment, but who have had such a ripping good time of it we don't give a fig and intend to keep on tinkering until we're so old and frail we can no longer be wheeled into our workspaces!

Long live tinkering and all the wonderful things to be learned from it! Long may you tinker with your kids! Now get to it!

A Academic Research On How Learning Works

This is in no way a comprehensive list of resources, but rather a list of work I personally have found relevant and illuminating. Much additional information in this area may be found with an Internet search.

Polin, L. (2002). Learning in Dialogue with a Practicing Community. In Duffy and Kirkley (Eds.), *Learner Centered Theory and Practice in Distance Education: Cases from Higher Education*, LEA Publishers.

Brown, A., Campione, J. (1994). Guided Discovery in a Community of Learners. In Kate McGilly (Ed.), *Classroom Lessons*, The MIT Press.

Brown, J.S., Collins, A., Duguid, P. (1989). Situated Cognition and the Culture of Learning. *Educational Researcher. Vol. 18. No. 1, pp. 32-42*.

Lave, J., Wenger, E. (1991). *Situated Learning: Legitimate Peripheral Participation*, Cambridge University Press.

Rogoff, B. (1994). Developing Understanding of the Idea of Communities of Learners. In *Mind, Culture and Activity*, Volume I, No. 4.

Rogoff, B., Matusov, E., White, C., (1996). Models of Teaching and Learning: Participation in a Community of Learners. In D. R. Olson and N. Torrance (Eds.), *The Handbook of Education and Human Development*, Oxford, UK: Blackwell.

Rogoff, B., Moore, L., Najafi, B., Dexter, A., Correa-Chávez, M., Solís, J. (2007). Children's Development of Cultural Repertoires through Participation in Everyday Routines and Practices. In J. E. Grusec and P. D. Hastings (Eds.), *Handbook of Socialization: Theory and Research*, pp. 490-515. New York, NY: The Guilford Press.

Bransford, J. D., Brown, A. L., Cocking, R. R., Editors (1999). *How People Learn: Brain, Mind, Experience, and School*. This is a massive report by the National Research Council (our tax dollars at work!) summarizing up to-date research on learning and pointing the way toward further research and best practices in teaching. It's available on the Web. Additional titles in the same vein were subsequently published, including *How Students Learn: Science in the Classroom* (2005).

B Evaluation Questionnaire for Students

We often give evaluation (*eval*) questionnaires to older students after class group sessions. We believe they help students think back over what they've just done and learned. Ideally, they also get practice writing and using new vocabulary. In our experience, few are truly interested in this sort of introspection and reanalysis, and many just want to continue tinkering until (a) the period is over or (b) we coerce them to sit down and do it (You're not taking your project home until you finish your eval!). There has rarely been a time when we've got kids in the free-form tinkering sessions to sit down and do this. I think the internal urge to self-evaluate must come with a bit of age.

We sometimes present the eval in the form of a journal that kids add to each week, and sometimes in the form of a paper we copy off and hand out. It also works just to write the prompts on the board and have kids write responses on their own paper. The eval often has four parts, and we encourage the students to focus on whichever part is the most interesting or useful for them.

1. "*Draw your project.*" This is important for kids who have trouble writing, and for visual/special kids.

2. "*What were the main steps you used in constructing your project?*" Ideally this gets the kids to think back about the process they used to learn how to do the project, but be forewarned that many kids are well content to remain ignorant about this. If they can focus, it is a good exercise just to write the trivial: cut the board, glued the dowel, attached the string, etc.

3. "*What did you learn by making your project?*" The biggie. More on this below.

4. *Content Questions.* These are like those in this book's project sections under *Check It Out*. They are questions designed to be answerable by means of playing around with the project, as opposed to quiz-type questions (What are the three states of matter?).

Question 3 seems straightforward; that's what we all want to know, right? Unfortunately the answers given rarely satisfy: "A lot." "More than I ever imagined." "Nothing."

In recent years, we've tweaked and kneaded that question to result in better answers. We then rotate through these prompts:

- Here is advice I would give to someone else who wants to build this project:
- This is the part of this project that I had trouble making and how I solved the problem:
- Something I would like to understand better about this project is:
- Some ways the project I built is different from the real-life version are:
- The part of this project that worked very well for me was:
- One thing I would do differently to improve this project if I did it again is:
- This project is related to science and engineering because:

Index

About the Author

Curt Gabrielson was the founding director of the Watsonville Environmental Science Workshop and works in science education in Timor-Leste. He has been a science educator for more than 20 years, with positions in the California public schools, the National University of Timor-Leste, and San Francisco's Exploratorium Teacher Institute.

CPSIA information can be obtained at www.ICGtesting.com
Printed in the USA
BVOW07s0131221015

423504BV00001B/1/P